Springer Theses

Recognizing Outstanding Ph.D. Research

For further volumes:
http://www.springer.com/series/8790

Aims and Scope

The series "Springer Theses" brings together a selection of the very best Ph.D. theses from around the world and across the physical sciences. Nominated and endorsed by two recognized specialists, each published volume has been selected for its scientific excellence and the high impact of its contents for the pertinent field of research. For greater accessibility to non-specialists, the published versions include an extended introduction, as well as a foreword by the student's supervisor explaining the special relevance of the work for the field. As a whole, the series will provide a valuable resource both for newcomers to the research fields described, and for other scientists seeking detailed background information on special questions. Finally, it provides an accredited documentation of the valuable contributions made by today's younger generation of scientists.

Theses are accepted into the series by invited nomination only and must fulfill all of the following criteria

- They must be written in good English.
- The topic should fall within the confines of Chemistry, Physics, Earth Sciences, Engineering and related interdisciplinary fields such as Materials, Nanoscience, Chemical Engineering, Complex Systems and Biophysics.
- The work reported in the thesis must represent a significant scientific advance.
- If the thesis includes previously published material, permission to reproduce this must be gained from the respective copyright holder.
- They must have been examined and passed during the 12 months prior to nomination.
- Each thesis should include a foreword by the supervisor outlining the significance of its content.
- The theses should have a clearly defined structure including an introduction accessible to scientists not expert in that particular field.

Rosa Córdoba Castillo

Functional Nanostructures Fabricated by Focused Electron/Ion Beam Induced Deposition

Doctoral Thesis accepted by
the University of Zaragoza, Spain

Springer

Author
Dr. Rosa Córdoba Castillo
Laboratorio de Microscopías Avanzadas
Instituto de Nanociencia de Aragón
Department of Condensed Matter Physics
Universidad de Zaragoza
Zaragoza
Spain

Supervisors
Prof. José M. De Teresa
Prof. Javier Sesé
Departamento de Fisica de la Materia Condensada
Universidad de Zaragoza
Zaragoza
Spain

ISSN 2190-5053 ISSN 2190-5061 (electronic)
ISBN 978-3-319-37553-3 ISBN 978-3-319-02081-5 (eBook)
DOI 10.1007/978-3-319-02081-5
Springer Cham Heidelberg New York Dordrecht London

Printed on acid-free paper

Springer is part of Springer Science+Business Media (www.springer.com)

Publications

1. R. Córdoba, J. Sesé, M. R. Ibarra, I. Guillamón, H. Suderow, S. Vieira, and J. M. De Teresa, "Non-local voltage in superconducting W-based nanowires grown by focused-ion-beam-induced-deposition", manuscript in preparation.
2. R. Córdoba, J. Sesé, T. I. Baturina, A. Y. Mironov, J. M. De Teresa, M. R. Ibarra, J. J. Palacios, I. Guillamón, H. Suderow, S. Vieira, M. Baklanov, and V. M. Vinokur, "Magnetic field-induced dissipation-free state in superconducting nanostructures", Nature Communications **4**: 1437 (2013).
3. R. Córdoba, J. Sesé, M. R. Ibarra, and J. M. De Teresa, "Autocatalytic growth of Co on pure Co surfaces using $Co_2(CO)_8$ precursor", Applied Surface Science **263**, 242–246 (2012).
4. R. Córdoba, R. Lavrijsen, A. Fernández-Pacheco, F. J. Schoenaker, T. H. Ellis, B. Barcones, J. T. Kohlepp, H. J. M. Swagten, B. Koopmans, M. R. Ibarra, J. J. L. Mulders and J. M. De Teresa, "Giant anomalous Hall effect in Fe-based microwires grown by focused-electron-beam-induced deposition", Journal of Physics D: Applied Physics **45**, 035001 (2012).
5. R. Córdoba, R. Fernández-Pacheco, A. Fernández-Pacheco, A. Gloter, C. Magén, O. Stéphan, M. R. Ibarra, J. M. De Teresa, "Nanoscale chemical and structural study of Co-based FEBID structures by STEM-EELS and HRTEM", Nanoscale Research Letters **6**, 1–6 (2011).
6. L. Serrano-Ramón, R. Córdoba, L. A. Rodríguez, C. Magén, E. Snoeck, C. Gatel, I. Serrano, M. R. Ibarra and J. M. De Teresa, "Ultra-Small Functional Ferromagnetic Nanostructures grown by Focused-Electron-Beam-Induced Deposition", ACS Nano **5**, 10, 7781–7787 (2011).
7. R. Lavrijsen, R. Córdoba, F. J. Schoenaker, T. H. Ellis, B. Barcones, J. T. Kohlepp, H. J. M. Swagten, B. Koopmans, J. M. De Teresa, C. Magén, M. R. Ibarra, P. Trompenaars, and J. J. L. Mulders, "Fe:O:C grown by focused-electron-beam-induced deposition; magnetic and electric properties", Nanotechnology **22**, 21, 025302 (2011).
8. R. Córdoba, J. Sesé, J. M. De Teresa, and M. R. Ibarra, "High-purity cobalt nanostructures grown by focused-electron-beam-induced deposition at low current", Microelectronic Engineering **87**, 1550–1553 (2010).
9. J. M. De Teresa, A. Fernández-Pacheco, R. Córdoba, J. Sesé, M. R. Ibarra, I. Guillamón, H. Suderow, S. Vieira, "Transport properties of superconducting amorphous W-based nanowires fabricated by focused-ion-beam-induced-deposition for applications in Nanotechnology", Mater. Res. Soc. Symp. Proc. **1181**, 1180-CC04-09 (2009).
10. J. M. De Teresa, R. Córdoba, A. Fernández-Pacheco, O. Montero, P. Strichovanec, M. R. Ibarra, "Origin of the difference in the resistivity of as-grown focused-ion- and electron-beam-induced Pt nanodeposits", Journal of Nanomaterials **2009**, 936863 (2009).
11. A. Fernández-Pacheco, J. M. De Teresa, R. Córdoba, M. R. Ibarra, "High-quality magnetic and transport properties of cobalt nanowires grown by

focused-electron-beam-induced deposition", Journal of Physics D: Applied Physics **42**, 055005 (2009).
12. A. Fernández-Pacheco, J. M. De Teresa, R. Córdoba, M. R. Ibarra, "Metal-insulator transition in Pt-C nanowires grown by focused-ion-beam-induced deposition", Physical Review B **79**, 17, 174204 (2009).
13. I. Guillamón, H. Suderow, S. Vieira, A. Fernández-Pacheco, J. Sesé, R. Córdoba, J. M. De Teresa and M. R. Ibarra, "Nanoscale superconducting properties of amorphous W-based deposits grown with focused-ion-beam", New Journal of Physics **10**, 093005 (2008).
14. A. Fernández-Pacheco, L. E. Serrano-Ramón, T. Tyliszczak, K. W. Chou, R. Córdoba, A. Szkudlarek, L. O'Brien, Cz. Kapusta, M. R. Ibarra, J. M. De Teresa, "Correlation between the magnetic imaging of cobalt nanoconstrictions and their magnetoresistance response", Nanotechnology **23**, 10, 105703 (2012).
15. S. Estrade, J. M. Rebled, M. G. Walls, F. de la Pena, C. Colliex, R. Córdoba, I. C. Infante, G. Herranz, F. Sanchez, J. Fontcuberta, F. Peiro,, "Effect of the capping on the local Mn oxidation state in buried (001) and (110) SrTiO3/La2/3Ca1/3MnO3 interfaces", Journal of Applied Physics **110**, 10, 103903 (2011).
16. F. J. Schoenaker, R. Córdoba, R. Fernández-Pacheco, C. Magén, O. Stephan, C. Zuriaga-Monroy, M. R. Ibarra, J. M. De Teresa, "Focused electron beam induced etching of titanium with XeF_2", Nanotechnology **22**, 26, 265304 (2011).
17. M. J. Martinez-Perez, J. Sese, F. Luis, R. Córdoba, D. Drung, T. Schurig, E. Bellido, R. de Miguel, C. Gomez-Moreno, A. Lostao, D. Ruiz-Molina, "Ultrasensitive broad band SQUID microsusceptometer for magnetic measurements at very low temperatures", IEEE Transactions on Applied Superconductivity **21**, 3, 345–348 (2011).
18. I. Guillamón, H. Suderow, S. Vieira, J. Sesé, R. Córdoba, J. M. De Teresa and M. R. Ibarra, "Direct observation of stress accumulation and relaxation in small bundles of superconducting vortices in tungsten thin films", Physical Review Letters **106**, 7, 077001 (2011).
19. D. Muñoz-Rojas, R. Córdoba, A. Fernández-Pacheco, J. M. De Teresa, G. Sauthier, J. Fraxedas, R. I. Walton, and N. Casañ, "High conductivity in hydrothermally grown $AgCuO_2$ single crystals verified using focused-ion-beam-deposited nanocontacts", Inorganic Chemistry **49**, 23, 10977–10983 (2010).
20. N. Marcano, S. Sangiao, M. Plaza, L. Perez, A. Fernández-Pacheco, R. Córdoba, M. C. Sanchez, L. Morellón, M. R. Ibarra, J. M. De Teresa, "Weak-antilocalization signatures in the magnetotransport properties of individual electrodeposited Bi Nanowires", Applied Physics Letters 96, 8, 082110 (2010).
21. A. Fernández-Pacheco, J. M. De Teresa, R. Córdoba, M. R. Ibarra, "Tunneling and anisotropic-tunneling magnetoresistance in iron nanoconstrictions fabricated by focused-ion-beam", Ion Beams and Nano-Engineering, Materials Research Society Symposium Proceedings **1181**, 35–40 (2010).

22. A. Fernández-Pacheco, J. M. De Teresa, A. Szkudlarek, R. Córdoba, M. R. Ibarra, D. Petit, L. O'Brien, H. T. Zeng, E. R. Lewis, D. E. Read, and R. P. Cowburn, "Magnetization reversal in individual cobalt micro- and nanowires grown by focused-electron-beam-induced-deposition", Nanotechnology **20**, 475704 (2009).
23. A. Fernández-Pacheco, J. M. De Teresa, R. Córdoba, M. R. Ibarra, D. Petit, D. E. Read, L. O'Brien, E. R. Lewis, H. T. Zeng, R. P. Cowburn, "Domain wall conduit behaviour in cobalt nanowires grown by Focused-Electron-Beam Induced Deposition", Applied Phys. Letters **94**, 192509 (2009).
24. M. J. Martínez-Pérez, J. Sesé, R. Córdoba, F. Luis, D. Drung, and T. Schurig, "Circuit edit of superconducting microcircuits", Superconductor Science and Technology **22**, 12, 125020 (2009).
25. I. Guillamón, H. Suderow, A. Fernández-Pacheco, J. Sesé, R. Córdoba, J. M. De Teresa, M. R. Ibarra and S. Vieira, "Direct observation of melting in a two-dimensional superconducting vortex lattice", Nature Physics **5**, 9, 651–655 (2009).
26. S. Estradè, J. M. Rebled, J. Arbiol, F. Peiró, I. C. Infante, G. Herranz, F. Sanchez, J. Fontcuberta, R. Córdoba, B. G. Mendis, A. L. Bleloch, "Effects of thickness on the cation segregation in epitaxial (001) and (110) La2/3Ca1/3MnO3 thin films", Applied Physics Letters **95**, 7, 072507 (2009).
27. J. V. Obona, J. M. De Teresa, R. Córdoba, A. Fernández-Pacheco, M. R. Ibarra, "Creation of stable nanoconstrictions in metallic thin films via progressive narrowing by focused-ion-beam technique and in situ control of resistance", Microelectronic Engineering **86**, 4-6, 639–641 (2009).
28. A. Fernández-Pacheco, J. M. De Teresa, R. Córdoba, M. R. Ibarra, "Exploring the conduction in atomic-sized metallic constrictions created by controlled ion etching", Nanotechnology **19**, 41, 415302 (2008).

Book Chapter

29. J. M. De Teresa, A. Fernández-Pacheco, R. Córdoba, M. R. Ibarra, Chapter 23 "Electrical transport properties of metallic nanowires and nanoconstrictions created by FIB", pp. 668–685; in Nanofabrication Using Focused Ion and Electron Beams: Principles and Applications (2012). Editors: P. E. Russell, I. Utke, S. Moshkalev, Oxford University Press. ISBN: 9780199734214.

Supervisors' Foreword

The use of focused electron and ion beams has certainly changed the way we image and pattern matter at small scales. Due to the high acceleration voltage in the kV range, such focused beams show short wavelengths due to the De Broglie relationship, and thus, are not limited by diffraction effects for practical purposes. This fact has encouraged researchers to build sophisticated instruments based on focused beams such as the Scanning Transmission Electron Microscope, with current imaging resolution below 0.1 nm. For micro- and nano-fabrication, the Scanning Electron Microscope (SEM), currently producing beam spots down to 1 nm size, and the Focused Ion Beam (FIB), currently producing beam spots down to 3 nm size, have become essential tools in Materials Science and Nanoscience laboratories. In particular, existing commercial or home-made equipment integrating FIB-SEM columns plus additional parts such as Gas Injection Systems (GIS) have become popular not only in the Semiconductor Industry for circuit edit and mask repair but also in research laboratories for many different applications like lamellae preparation, cross-sectional imaging, subtractive nanopatterning, establishment of electrical connections to nanostructures, in-situ nanomechanical studies, the growth of functional optical, magnetic, conductive, insulating, semiconductor and superconductor nanostructures, etc. Thus, several one-step lithography processes can be done in a FIB-SEM without the painful use of resist-based processing, opening new routes for nanofabrication.

Rosa Córdoba's Ph.D. thesis exploits the use of FIB-SEM equipment including GIS for the growth of tailored magnetic and superconducting nanostructures. The used techniques are coined Focused Electron Beam Induced Deposition (FEBID) and Focused Ion Beam Induced Deposition (FIBID). In her thesis, Dr. Córdoba has produced and characterized high-quality ferromagnetic nanostructures grown by FEBID and using $Co_2(CO)_8$ and $Fe_2(CO)_9$ as precursor molecules. By changing growth parameters such as beam current and voltage, flux pressure, dwell time, substrate temperature, etc., she has been able to achieve Co magnetic deposits with high metal content (>95 %) and high resolution (30 nm). Those deposits were subsequently found to be functional from the magnetic point of view, with great potential for their integration in magnetic sensors, memories, and logic. The microstructure of these cobalt deposits was thoroughly characterized by her and the role of autocatalytic effects during certain growth conditions unveiled. On the other hand, the Fe magnetic deposits could not be grown with such purity but showed interesting magnetotransport properties such as a Giant Anomalous Hall Effect. The second part of the thesis focused on the growth and characterization of nanostructured grown by FIBID and using $W(CO)_6$ as precursor molecules. In 2004 it was discovered that the resulting material is a superconductor with a critical temperature around 5 K. The important applications of the FIBID technique in the semiconductor industry were then also applicable for superconducting electronic devices. This thesis includes a complete characterization of this new superconductor, and the influence of the growing conditions. It is shown that the material is amorphous, presenting very low pinning for the superconducting vortices and excellent characteristics for studying the vortex lattice using Scanning Tunneling Microscopy. Then, she explored the limits of the FIBID technique and could fabricate the narrowest, but still continuous, lines of 50 nm. This achievement has permitted the observation of fascinating new size effects in the dynamics of vortices that appeared during the magnetotransport experiments presented in this thesis.

Zaragoza, July 2013

Prof. José M. De Teresa
Prof. Javier Sesé

Contents

Acronyms

AFM	Atomic force microscopy
AHE	Anomalous Hall effect
AMR	Anisotropic magnetoresistance
BCS	Bardeen–Cooper–Schrieffer
BE	Binding energy
BSE	Backscattered electron
CASINO	Monte Carlo simulation of electrons in solids
CCD	Charge-coupled device
CDEM	Continuous-dynode electron multiplier
CNT	Carbon nanotube
CVD	Chemical vapor deposition
DC	Direct current
DOS	Density of states
DT	Dwell time
EBL	Electron beam lithography
EDS	Energy dispersive X-ray spectroscopy
EELS	Electron energy loss spectroscopy
ELNES	Energy-loss near-edge structure
ES-VRH	Efros–Shklovskii variable range hopping
ETD	Everhart–Thornley detector
fcc	Face-centered cubic
FEB	Focused electron beam
FEBID	Focused electron beam-induced deposition
FEBIE	Focused electron beam-induced etching
FEG	Field emission gun
FFT	Fast Fourier transform
FIB	Focused ion beam
FIBID	Focused ion beam-induced deposition
FIBIE	Focused ion beam-induced etching
GAHE	Giant anomalous Hall effect
GIS	Gas injection system
GMR	Giant magnetoresistance
HAADF	High angle annular dark field
hcp	Hexagonal close packed

HDD	Hard disk drive
HE	Hall effect
HRTEM	High-resolution transmission electron microscopy
HV	High vacuum
ITMR	Intergrain tunneling magnetoresistance
LAMH	Langer–Ambegaokar–McCumber–Halperin
LMIS	Liquid metal ion source
MBE	Molecular beam epitaxy
MBJ	Mechanical break junction
MEMS	Microelectromechanical systems
MFM	Magnetic force microscopy
MIT	Metal–insulator transition
MOKE	Magneto-optic Kerr effect
MR	Magnetoresistance
MRAM	Magnetic random access memory
M_s	Saturation magnetization
MTJ	Magnetic tunnel junction
NEMS	Nanoelectromechanical systems
NW	Nanowire
OHE	Ordinary Hall effect
PCA	Principal component analysis
PE	Primary electron
PLD	Pulsed laser deposition
PPMS	Physical properties measurement system
PS	Phase slip
QPSC	Quantum phase slip center
RAM	Random access memory
RGA	Residual gas analyzer
RIE	Reactive ion etching
RRR	Residual resistivity ratio
RT	Refresh time
SC	Superconductor
SE	Secondary electron
SEM	Scanning electron microscopy
SI	Secondary ion
SQB	Superconducting quantum bit
SQUID	Superconducting quantum interference device
SRIM	Stopping and range of ions in matter
SSPD	Superconducting single-photon detector
STEM	Scanning transmission electron microscopy
STM	Scanning tunneling microscopy
STS	Scanning tunneling spectroscopy
TAPS	Thermally activated phase slip
TEM	Transmission electron microscopy
TLD	Through-the-lens detector

TMR	Tunneling magnetoresistance
TRIM	Transport of ions in matter
UHV	Ultrahigh vacuum
VD	Volume per dose
VRH	Variable range hopping

Chapter 1
Introduction

This chapter we will describe the concepts of nanotechnology and nanoelectronics, then we will go on to the general fundamentals of the fabrication of nanostructures by FEBID and FIBID techniques with potential applications in superconductivity and spintronics. This is the general framework within which this thesis is classified, which is devoted largely to the fabrication and characterization of nanostructures that exhibit ferromagnetic or superconducting properties.

Therefore, we will address the principles of the nanofabrication techniques that we have used, FEBID and FIBID. In addition, we will introduce the physical phenomena associated with ferromagnetic nanostructures studied, such as magnetoresistance and the anomalous Hall effect, continuing with the phenomena associated with superconducting nanostructures, such as the local and nonlocal vortex motion, among others.

1.1 Nanotechnology

Nanotechnology deals with the study and control of matter on a small scale, from 1 to 100 nm, the nanometer being 1×10^{-9} m, equivalent to about five atoms in line (see Fig. 1.1).

In 1959, Richard Feynman (Nobel Prize in Physics 1965) predicted according to the known laws of physics, that atoms would be studied and manipulated individually [1]. It seems that, once again, the dream of a man motivated and encouraged the scientific community to develop the tools for that prediction to come true, and it is clear that nanotechnology has revolutionized our industries and our lifes since the late 20th century.

Nanoscience and nanotechnology are multidisciplinary fields of science, bringing together disciplines such as physics, chemistry, engineering, biology and medicine. Therefore, the impact and implications of findings generated from nanotechnology can have incalculable dimensions, both economically and socially. We highlight the developments and applications in medicine by diagnosis, treatment and prevention; in energy by nanoparticles; in electronics by magnetic

R. Córdoba Castillo, *Functional Nanostructures Fabricated by Focused Electron/Ion Beam Induced Deposition*, Springer Theses, DOI: 10.1007/978-3-319-02081-5_1,

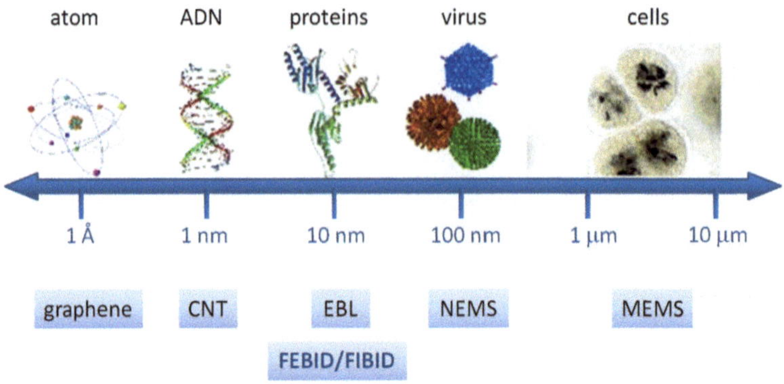

Fig. 1.1 Sketch illustrating the metric scale that applies in nanoscience and nanotechnology. It ranges from MEMS size (microelectromechanical systems), via NEMS (nanoelectromechanical systems), the resolution in EBL (electron beam lithography), FEBID (focused electron beam induced deposition), FIBID (focused ion beam induced deposition), and CNT size (carbon nanotubes) to graphene, a monolayer of carbon atoms

memories for computers, electronic devices, thin films; in textiles with the fabrication of nanocomposites; in construction using new materials, coatings, CNTs; in engineering, chemistry and biotechnology by using microreactors, biosensors, and nanostructured catalysts (see Fig. 1.2).

Materials at the nanoscale can show different properties compared to those exhibited at the macroscale. This fact allows us to study fundamental physical phenomena, such as quantum size effects, which enable unique applications in nanoelectronics.

Fig. 1.2 The most important application fields of nanoscience and nanotechnology

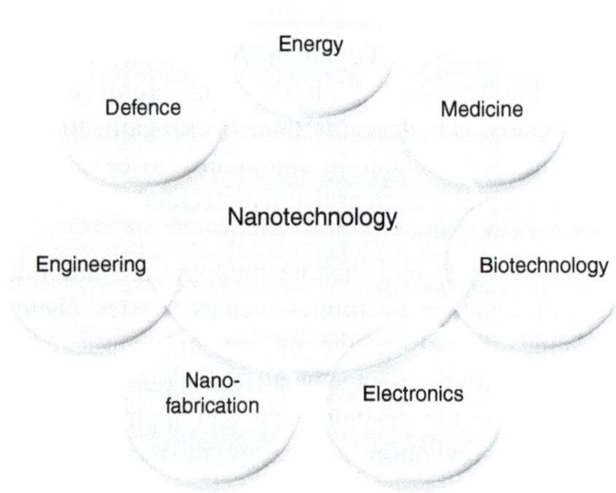

1.2 Nanoelectronics

The data storage industry generates billions of dollars a year in the USA, with hundreds of millions of electronic devices. Magnetic data storage is widely used in applications that are very common today, such as hard disks of computers, non-volatile memories, credit cards and other electronic devices [2].

The study of magnetic materials and their properties will lead to improvements in information storage, fabrication of sensors, and other devices, as has occurred in recent decades with the discovery of anisotropic magnetoresistance (AMR), giant magnetoresistance (GMR) and tunneling magnetoresistance (TMR). New concepts based on the magnetism of materials, such as nucleation and/or propagation of the domain wall will be useful in magnetic logic, processing information through magnetic circuits, which are part of these new transistors [3].

The emergence of new materials with excellent electrical properties, such as graphene, whose discovery in 2004 [4, 5] led to the Nobel Prize in Physics for A. Geim and K. S. Novoselov in 2010 [6], is now a new revolution in micro- and nanoelectronics [7].

1.3 Nanofabrication by Means of a "Dual Beam" System

1.3.1 General Description of a "Dual Beam" System

The dual beam microscope is made up of a focused electron beam (FEB), which is also called a scanning electron microscope (SEM), and a focused ion beam (FIB).

Obviously, the main parts of the dual beam microscope are the electron and ion columns. The electron column is located on the vertical of the vacuum chamber, and the ion column 52° from the electron column. The vacuum system used in the electron column comprises two ionic pumps, one for the electron source, which is at the top of the column at a pressure of $\sim 2.5 \times 10^{-10}$ mbar, and another for the rest of the column at a pressure of $\sim 2 \times 10^{-7}$ mbar. The ion column has one ionic pump with a pressure of $\sim 2 \times 10^{-7}$ mbar. Moreover, the vacuum system in the process chamber consists of a rotary pump and a turbomolecular pump operating at high vacuum; the chamber base pressure is around $\sim 1 \times 10^{-6}$ mbar. The position and movement of the specimen in the *x*-, *y*-, and *z*-directions and rotation and tilt into the process chamber is controlled by a motorized stage, positioning the sample down to the FEB and the FIB. It may take images of the same object simultaneously with both beams. Different types of detectors to collect secondary electrons (SEs), backscattered electrons (BSEs) and secondary ions (SIs) emitted by the sample are used to create an image. In Fig. 1.3 we can see a diagram of the most important parts of a dual beam system.

Micro- and nanofabrication using a dual beam system has several advantages over other microfabrication techniques such as photolithography. Direct

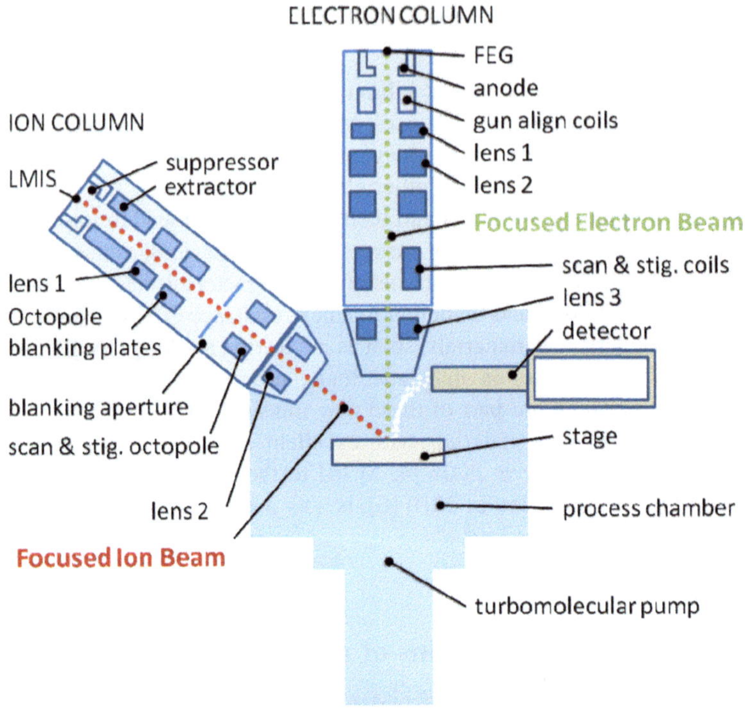

Fig. 1.3 Sketch of a "Dual Beam" system

fabrication of patterns at nanometer scale is possible without the use of masks or resists. Furthermore, it is possible to achieve a resolution of 3 nm [8] and is very versatile for a wide range of materials and geometries. Some applications of this system have been developed and integrated in the semiconductor industry, such as the fabrication of integrated circuits, for the repair of masks used in photolithographic processes [9], circuit editing, and the fabrication of samples used in transmission electron microscopy [10, 11].

1.3.1.1 Focused Electron Beam

The electron column is based on a field emission gun (FEG). The electrons are extracted from a tungsten filament by applying a bias voltage from 200 V to 30 kV. The accelerated electron beam leaves the source and is focused by the condenser and objective lenses. Thus is achieved to obtain an image with a resolution of ~0.9 nm.

In the electron irradiation, energetic electrons interact with the atoms of the substrate. In Fig. 1.4(a) we show the most relevant physical processes that electron irradiation involves, such as the SEs, BSEs and X-ray signals, and in Fig. 1.4(c)

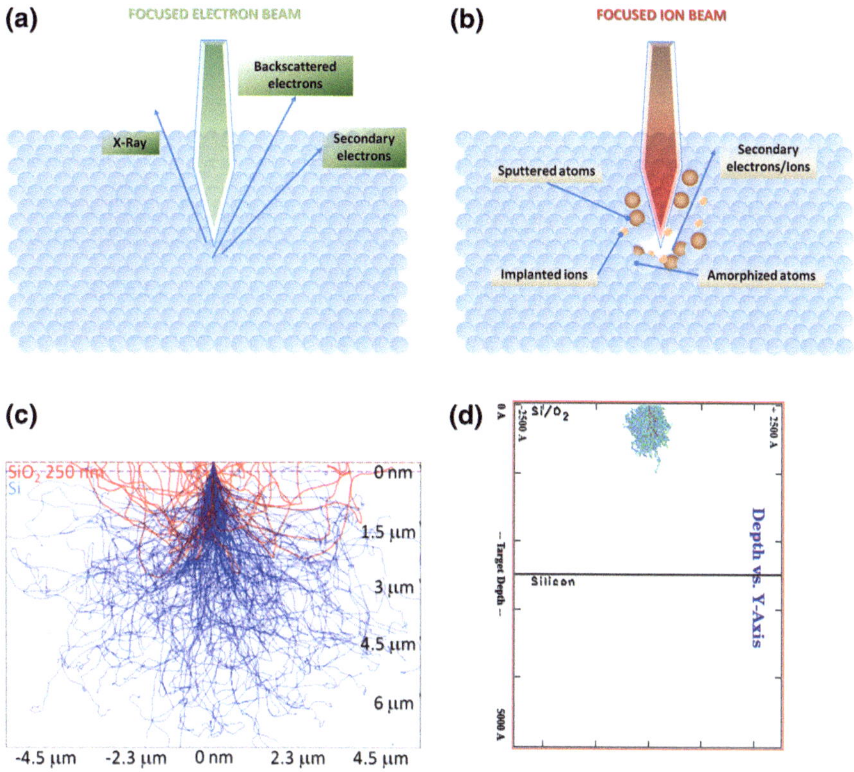

Fig. 1.4 Sketch of the most important interactions that occur between the FEB (**a**) or FIB (**b**) and solid matter. **c** Simulation of the FEB interactions at 30 kV with incidence angle perpendicular to the substrate surface by the CASINO software. **d** Simulation as in (**c**) but using the Ga^+ FIB and performed using SRIM software

we show a simulation performed using the CASINO software ("*monte CArlo SImulation of electroN trajectory in sOlids*") [12, 13], which indicates that ~6 µm is the length of the interaction volume of the FEB at 30 kV with an incidence angle perpendicular to the substrate (silicon oxidized thermally, 250 nm SiO_2).

1.3.1.2 Focused Ion Beam

The ion column is based on a liquid metal ion source (LMIS) [14]. The ions are extracted by an electric field from the liquid metal layer, in this case Ga^+, which covers a sharp tungsten tip. The beam is focused to ~5 nm diameter by a suitable optical system and can operate using an acceleration voltage from 0.5 to 30 kV. The FIB operates similarly to the SEM, but using an ion beam instead of an electron beam (Fig. 1.5), which allows us to take images, to sputter material and to grow material.

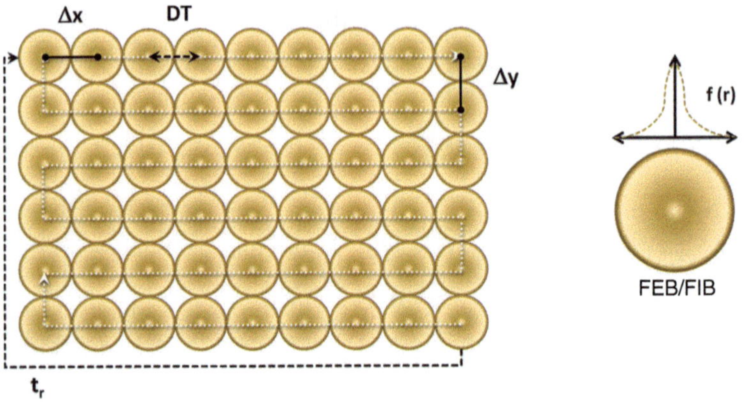

Fig. 1.5 Scheme of both FEB and FIB scan with serpentine shape (*bottom to top*). The beam has a Gaussian profile, the time the beam remains at each point is *DT*, the beam movement from one point to another along the *x*- and *y*-axes are Δx and Δy, respectively. The pattern scan is repeated for the number of passes after the refresh time, *RT*

In Fig. 1.4b we represent the most important interactions between the FIB and solid matter. Also, in (d) we show a graph from a simulation performed with the TRIM software, indicating that the penetration depth of the Ga^+ FIB, accelerated at 30 kV and incident perpendicular to the substrate is $\sim$100 nm. TRIM/SRIM (Transport of Ions in Matter/Stopping and Range of Ions in Matter) [15] is the standard software to evaluate the interactions of accelerated ions in matter.

1.3.2 Focused Ion Beam Milling

Metaphorically we could talk about FIB as a full box of brushes of different diameters available to an artist. So, we can draw thick or thin strokes with a nanometer precision on a canvas, depending on its diameter, defining lines, areas and patterns, which finally form a design. Thus, if we put aside the functionality of these patterns, the milling of structures by FIB could be considered as art at the nanoscale. FIB interactions with matter occur from the kinetic energy loss of the ions. Therefore, if the ion cannot scatter outside the substrate surface will be implanted at some depth from the substrate surface. Some examples of these "pieces of art" are shown in Fig. 1.6, where we find images of logos that have been fabricated by FIB as a symbol of gratitude to the institutions that have allowed me the use of infrastructures and equipment for experimental work for this thesis: (a) SEM image of the University of Zaragoza logo on a Si substrate, (b) FIB image of the Advanced Microscopy Laboratory logo on a Cu substrate and (c) FIB image of the Institute of Nanoscience of Aragon logo on a Cu substrate.

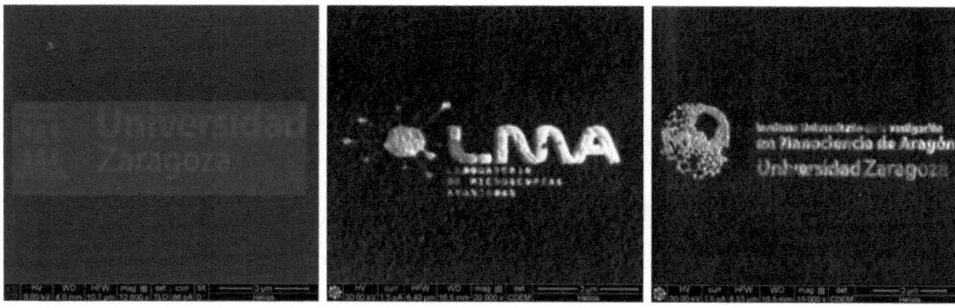

Fig. 1.6 Images of the logos milling by FIB (*left to right*) of the University of Zaragoza, the Advanced Microscopy Laboratory and Institute of Nanoscience of Aragon

From a functional point of view, the milling of nanostructures by FIB allows a new approach to micro- and nanoelectronic systems, such as the design or modification of nanomechanical devices that are part of transistors [16, 17], the milling of nanostructures [18–20], nanofabrication in three dimensions [11] and the nanofabrication of AFM tips [21, 22].

1.3.3 Focused Electron/Ion Beam Induced Etching

Focused electron/ion beam induced etching, FEBIE and FIBIE respectively, are techniques similar to the well-known reactive ion etching (RIE). The difference is that the etching is local in FEBIE/FIBIE, not global. The differences between the FEB and FIB obviously have a great influence on the etching results when using gas etching. The etching mechanism by gases may be due to one or more effects. The gases can be adsorbed on the sample surface and react directly with the sample surface, without the beam assistance to activate the reaction; or they can be activated by the secondary electrons generated by the primary beam. Adsorbed gas molecules can react with the material surface and attack it easily. Gases can also react with the pulverized material near to the surface to prevent redeposition [23].

The use of gas etching combined with a FIB is useful to increase the etching rate at low currents, and besides the sputtered material reacts, forming a volatile compound and preventing redeposition. For example, I_2 reacts with Al and W FIBID faster than that of SiO_2, which thus allows a contact to be cut, minimizing the etching in the insulating material [24].

The FEB does not produce a physical sputtering effect at the acceleration voltages used in a conventional SEM. The etching processes are based on the reactivity of adsorbed gas with the substrate and the volatility of reaction products. The process has been demonstrated in materials such as SiO_2, SiN, Al, Cr, CrO_x, TaN, GaAs and Ge [25–32] using XeF_2, NOCl, or Cl_2 as etching gases. Furthermore, Ti is susceptible to etching by FEBIE [33], see Fig. 1.7, as well as by FIBIE [34] using XeF_2 as the etching gas.

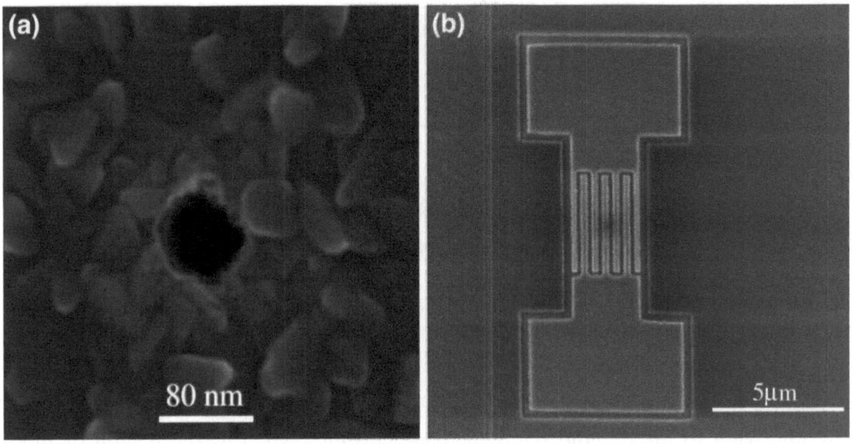

Fig. 1.7 FEBIE on a thin Ti film of 200 nm thickness by XeF_2: **a** SEM image of a hole with a circle shape (80 nm in diameter) fabricated by FEBIE. **b** Interdigitated nanoelectrodes with a resolution of ~200 nm. Reprinted with permission from [33]. Copyright (2011), IOP

1.3.4 Focused Electron Beam Induced Deposition

Focused electron beam induced deposition (FEBID) is a technique developed in the 1980s [35–37], which uses a focused electron beam, usually from SEM/FEB, to dissociate a precursor material, creating a selective deposit. The main advantages of this technique compared to traditional ones are: (1) deposition on any substrate, in any place where it is possible to take a FEB image, (2) one-step process, (3) the growth of three-dimensional structures, and (4) control of their thickness and lateral size at the nanoscale. The deposition mechanism can be considered as the well-known chemical vapor deposition (CVD) assisted by a focused electron beam, usually using organometallic compounds as precursor materials, see Fig. 1.8(a). The best resolution achieved by this technique was obtained by van Kouwen et al. [8], where Pt dots of 3 nm size were grown. These techniques, developed in the last 30 years, offer some exciting prospects in a wide range of applications in the field of micro- and nanoelectronics systems. More detailed information about the FEBID process can be found in the following reviews published by different authors [23, 38–40].

Going on to the FEBID process, first we must take into account the interactions that occur, such as those which occur between precursor molecules and the substrate, between the FEB and the substrate, and between the electrons generated in the substrate and the precursor molecules. The adsorption of precursor molecules on the substrate surface can be accomplished by chemisorption or physisorption; the residence time is an important parameter to consider. The probability that the FEB will dissociate precursor molecules is high at long residence times. However, when the deposition process begins, the interface is altered, changing from the

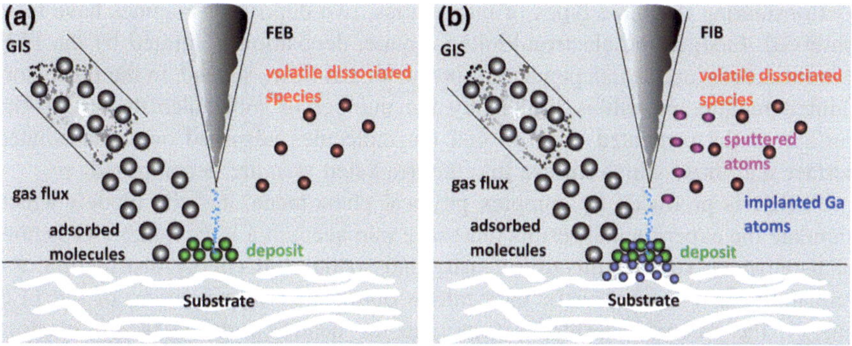

Fig. 1.8 Sketch of (**a**) focused electron beam induced deposition and (**b**) focused ion beam induced deposition. (GIS: gas injection system)

substrate surface to the deposit surface. A study that presents results on the reactions that take place between adsorbed molecules on the substrate surface and the FEB at low temperatures (<100 K) and in ultrahigh vacuum systems ($\sim 10^{-8}$ mbar) has been published recently [41].

As we saw in Fig. 1.4(a), the FEB produces several interactions in the substrate. First, the primary beam collides with the solid and is deflected from its initial trajectory. If the collision is inelastic, part of the initial energy can be transferred from the primary beam to other electrons in the solid. These electrons interact with the solid and disperse. The low-energy electrons generated, with energy less than 50 eV are called SEs; the high energy electrons are called BSEs. As mentioned previously, the FEBID process is more complex because the SEs and BSEs are generated on a substrate that is modified during the process itself. The most obvious example is when we grow pillars, because the electrons that contribute to growth are confined to it, and also because the FEB can generate a rise in temperature owing to the decrease of the possibility of heat being dissipated by the substrate [42].

The most important interaction between the FEB and precursor molecules is the decomposition of these, but we must not forget that the FEB can also stimulate desorption of precursor molecules. The probability that an electron induces the splitting of a precursor molecule bond depends on its bond energy, and is expressed as a cross section $\sigma_{(E)}$ [cm^2]. The higher $\sigma_{(E)}$, the greater the probability that the molecule bonds dissociate. However, it is a probability that depends on many parameters, such as the binding energies of the molecule and the specific geometry of the molecule adsorption. Low-energy electrons with energy of the order of tens of electron volts can induce dissociation of the bonds of precursor molecules, in the same way as they cause dissociation of contaminant species such as hydrocarbons found within the process chamber during the FEB scan. Thus, low-energy electrons, such as SEs, play a principal role in the process of dissociation of the precursor molecules and therefore the deposit growth. However, we cannot exclude from this process the contribution due to the high-energy electrons, such as the primary beam or BSEs.

Considering the three types of interactions, two deposition regimes have been observed. First, in the electron-limited regime, deposition is limited by the FEB current density and is independent of precursor gas flow. Second, in the precursor-limited regime, growth is limited by the number of molecules that reach the surface of the irradiated area, so that the molecules adsorbed on the irradiated surface and those which diffuse into the irradiated area are decomposed.

FEBID is produced by complex physical phenomena, and the models which simulate the experimental results must take into account a large number of factors or parameters, such as electron density and spatial and energy distribution, the probability that the precursor molecule is dissociated as a function of the FEB energy, the precursor residence time, and the heating induced by the electron beam. Simulations based on the Monte Carlo model allow us to study the spatial distribution and the energy spectrum of electrons generated in a solid, and thus become a very useful tool [42].

The deposition rate is defined by

$$R = \int_{0}^{E_0} f(E)\sigma_{diss}(E)NdE \tag{1.1}$$

where E_0 is the energy of the primary beam, $f(E)$ the electron flow, $\sigma_{\text{diss}}(E)$ the probability that the precursor molecule bond dissociates at a given energy, and N the molecular density at the substrate surface. Some authors found experimentally a maximum in $\sigma_{\text{diss}}(E)$ at an energy value of a few tens of electron volts [43]. These values are very close to the position of the maximum of the SE distribution generated by the FEB, which thus leads to a high dissociation of precursor molecules.

However, the decomposition of the precursor material is not complete owing to their nature, their binding energies and their probability of dissociation under the working conditions of conventional electron microscopy. These working conditions are room temperature in the process chamber, high vacuum (in the range of 10^{-6} mbar) in the absence of reactive gases and, on the other hand, the low generation of secondary electrons at typical working voltages (1–30 kV). The latter (secondary electrons) are mainly responsible for the decomposition process of the precursor material and as a consequence the deposition of materials.

Thus, the deposit comprises the desirable element, which is usually metallic, plus carbon and oxygen from the precursor material and contaminant species (hydrocarbons, oxygen and water vapor) present in the vacuum chamber, which in general impair the functionality of the nanostructures. In the literature, we find the deposition of nanostructures of different types of materials such as Pt, Pd, Cr, W, Co, and Fe with a metal content <50 % [23, 39, 40, 44]. In Fig. 1.9 we show a SEM image of Pt Christmas decorations grown by FEBID. Different strategies have focused on the purification of these deposits to raise the metal content and thus the physical properties of them have been studied in recent years [43–47].

Fig. 1.9 SEM image of Pt Christmas decorations fabricated by FEBID on a Si substrate

1.3.5 Focused Ion Beam Induced Deposition

Focused ion beam induced deposition (FIBID) is an analogous technique to FEBID but uses a focused ion beam instead of a FEB, see Fig. 1.8(b). Therefore, the comments in Sect. 1.3.4 apply to this as well [10, 40]. However, we must not forget that the action of impact and sputtering of the FIB has not disappeared. In the balance of processes, this contribution must be overcome by the deposition rate. To achieve a suitable deposition we select the FIB current according to the material volume to be grown, because if the FIB current density is too high the substrate surface will be sputtered instead of growing a deposit, whereas, if the FIB current density is too low, the deposition rate will be low and the material deposited will contain a high concentration of material precursor that has not decomposed.

There are several secondary effects that influence the fabrication process by FIBID, such as ion implantation, heat local generation, redeposition of material, SE effects and deposition due to scattered ions. In fabrication processes a heavy ion FIB such as Ga^+ is usually used, in which the scattered ions can have a significant influence on the growth of nanostructures compared with the effects produced by the secondary electrons. The deposit generated by scattered ions can grow at distances of a few micrometers from the base of the nanostructures, and depends on the energy and primary ion flux used in the deposition process. Therefore, the halo in the deposits generated by FIBID is largely due to the dissociation of precursor molecules by scattered Ga^+. The shape and the thickness of the halo are directly proportional to the scattered Ga^+ ion flow and the scattered ion intensity close to the base of the structure.

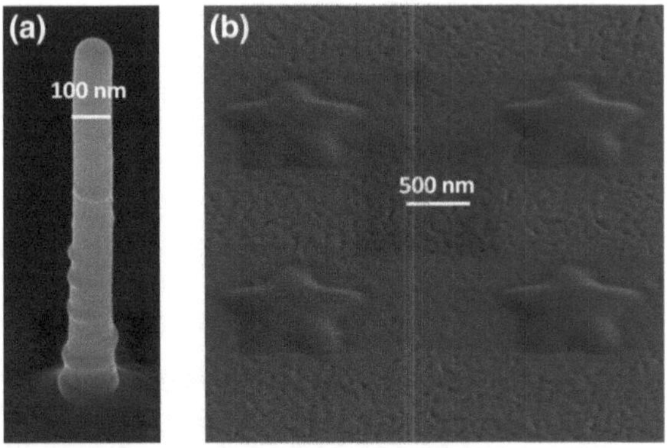

Fig. 1.10 SEM images of W FIBID structures: **a** a pillar and (**b**) stars (images tilted at 52°)

In addition, other phenomena may occur, such as sputtering and decomposition of the precursor molecules by SEs. Thus, in areas very close to the structure, the growth profile seems quite complex. However, the influence of these phenomena may be negligible for relatively long distances, more than 200 nm away from the base of the deposit [48].

In the literature we can find many studies to support the idea that the growth of nanostructures by FIBID could play a crucial role in many applications in nanotechnology. As examples, we can enumerate the growth of three-dimensional structures—either W, as shown in Fig. 1.10(a), or Pt (Fig. 1.11)—which could be used to contact nanoparticles, nanowires, etc. [49–54]; the deposition of structures with many shapes, as we can see in Fig. 1.10(b); the use of SiO_2 deposits for sealing MEMS devices (under vacuum) [55]; Pt deposits grown by FIBID as protective films in the fabrication of transmission electron microscopy (TEM) samples by FIB; repair of integrated circuits using W FIBID [24, 56], among others.

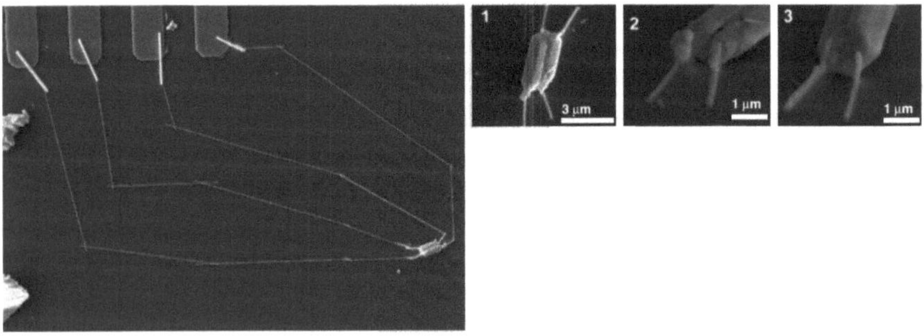

Fig. 1.11 SEM images of Pt deposits grown by FIBID that contact a particle, thus allowing its electrical properties to be studied. Adapted with permission from [54]. Copyright (2010), ACS

1.3.6 Comparison of FEBID and FIBID Techniques

The nanostructures studied in this thesis were fabricated by these two techniques. In Table 1.1 we summarize the main differences between them.

1.3.7 Comparison of FEBID/FIBID and Electron Beam Lithography

In Fig. 1.12 we summarize the most significant advantages of the nanofabrication techniques FIBID and FEBID compared to more conventional techniques such as EBL.

After discussing the versatility and the many advantages that these new techniques exhibit, we must be aware of the large number of parameters to be studied and optimized for best results: deposit dimensions (*x*,*y*,*z*), beam acceleration voltage, beam current, dwell time, pixel, refresh time, precursor gas flow, chamber

Table 1.1 Main differences between FEBID and FIBID techniques

FEBID	FIBID
Low deposition rate	High deposition rate
Halo generation	Halo generation
–	Sputtering of the substrate
–	Ga^+ ion implantation
–	Substrate amorphization
Low metal content, with exceptions	High metal content
–	Ions could affect material properties
Lateral size of deposits <3 nm	Lateral size of deposits <10 nm

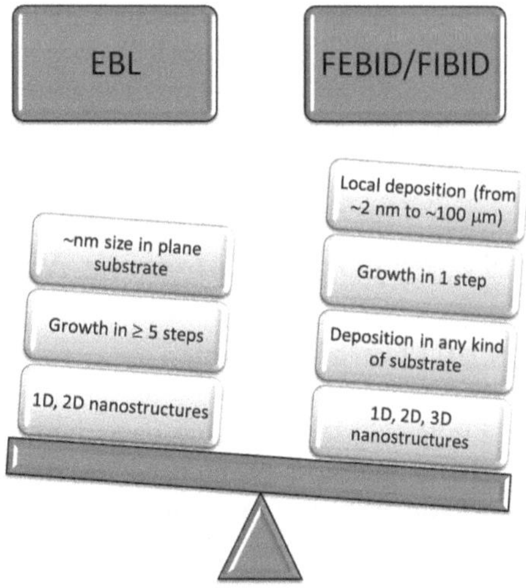

Fig. 1.12 Comparison diagram as a balance of the virtues of the techniques FEBID/FIBID compared to those of EBL

base pressure, nature and cleaning of the substrate, substrate temperature, the position of the precursor gas needle with respect to the area of interest of the sample, number of scans, scanning strategy. We must not forget the drawbacks or deficiencies that may arise. In the case of FEBID: halo, low metal content $\sim$30 at % (except for Fe content >80 % and Co >90 %), high resistivity (several orders of magnitude compared to the pure metal, except Co and Fe, as we will show in Chaps. 3 and 4 of this thesis) and low volume per dose (from 0.01 to 0.0005 $\mu m^3/nC$). In the case of FIBID: halo, Ga^+ implantation, FIB induces amorphization, disorder (in the case of W deposits this is an advantage, as we will see in Chap. 5) and sputtering of the first nanometers of the substrate.

1.4 Nanostructures Studied in This Thesis

Today, nanostructures can be fabricated using conventional techniques that have also been introduced for nanotechnology, such as MBE (molecular beam epitaxy), PLD (pulsed laser deposition), sputtering, or CVD or combined with EBL, etc. However, the advantages put forward in previous sections led us to the fabrication and study of functional nanostructures using other, more recent and less exploited techniques. The nanostructures fabricated in this work were grown using FIBID and FEBID techniques widely documented in the previous sections.

1.5 Magnetic Nanostructures

1.5.1 Introduction to Spintronics

Spintronics is an emerging technology that uses the phenomenon of spin transport in data magnetic storage devices, magnetic logic, magnetic sensors, etc. Thus, in recent years a new generation of spintronic devices based on new materials on the nanoscale has been developed.

1.5.2 Properties of Magnetic Nanostructures

1.5.2.1 Electric Conduction

The following sections focus on an introduction to the magnetic properties and the electrical transport phenomena associated with granular metals, such systems being similar in nature to the magnetic structures (Co and Fe) grown by FEBID. Granular metals consist of a mixture of metallic and non-metallic components;

when the composition of this disordered system changes, their electrical properties do as well. We distinguish three electrical conduction regimes: metal, metal–insulator transition and insulating or semi-insulating.

Metallic regime: In the metallic regime, the electrical resistivity is influenced by electron–phonon interactions of the metal structure. At high temperatures, these interactions are responsible for the increase in resistivity. Further, metals have a positive residual resistivity ratio (RRR), defined as RRR = $R(T)/R(0\ \text{K})$.

Metal–insulator transition (MIT) regime: The metal–insulator transition indicates that the granular materials go from metallic to insulator with decreasing metal content, but in any case pure metallic or insulating behavior is observed. Metallic conduction occurs in materials with high metal concentration, which percolate, creating conduction paths. The percolation paths decrease at low metal content, and the electron activation between metallic islands begins to be significant for the conduction mechanism.

Insulating or semi-insulating regime: The conductivity is in the insulating or semi-insulating regime in granular systems when the metallic islands are completely surrounded by an insulating matrix. Then, the electrical conduction is mainly based on the tunneling or hopping of electrons from island to island.

The conductivity in this regime can be explained by a standard model [57] based on two scenarios: first, the electron transport is due to tunneling between close neighbors and it is thermally activated; second, an electron requires an amount of energy to jump to the charged island to overcome the additional electrostatic energy. For disordered granular systems, it was found that the conductivity σ should behave as

$$\sigma \propto \exp(-(T_0/T)^x) \tag{1.2}$$

where T_0 and $x \sim 0.5$ are constants [58]. This indicates that the conduction pathways vary depending on the temperature, and also produces a change in the balance between the hopping and the activation energy. This mechanism is known as variable range hopping (VRH).

In Fig. 1.13 we represent some illustrations that differentiate between various types of electrical conduction as a function of the composition and microstructure of a given system. In (a), the metal concentration is low, small metal particles form

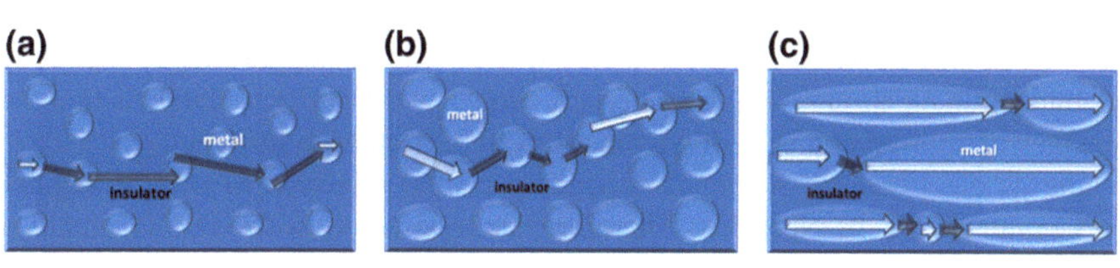

Fig. 1.13 Sketches showing various types of conduction in granular systems depending on the metal content and microstructure. **a** Tunnel conduction, semi-insulating behavior or insulating, low metal content. **b** Tunnel conduction or hopping with some metallic paths, MIT. **c** Metallic conduction

isolated islands embedded in an insulating matrix. The electrical conductivity is relatively small and requires thermal activation. A mechanism of tunnel conduction or hopping between the metal particles can be concluded. In (b), the concentration of metal in the insulating matrix increases, although there is still a semi-insulating structure close to the metal–insulator transition. In (c), the metal concentration is high and the microstructure consists of a metallic matrix with small insulator inclusions or islands. Therefore, this system shows purely metallic conduction.

1.5.2.2 Magnetoresistance

The magnetoresistance (MR) is the material property that changes the electrical resistance when an external magnetic field is applied and is defined as the ratio of the change in resistance at the applied field (compared to zero field) to the resistance at zero field

$$MR(\%) = 100\,\frac{\rho(H) - \rho(H=0)}{\rho(H=0)} \tag{1.3}$$

This phenomenon has been widely studied in a large number of investigations of great interest, with the result that today many magnetoresistive materials are available in commercial technology, such as magnetic sensors, magnetic recording heads and storage devices, which are based on phenomena such as giant magnetoresistance (GMR), anisotropic MR (AMR) or tunneling MR (TMR), see Table 1.2.

GMR was discovered in 1988 by Albert Fert [59], a fascinating discovery for which he was awarded the Nobel Prize in Physics in 2007 together with Peter Grünberg. Its origin lies in the spin-dependent scattering of electrons that takes place in multilayer or granular materials (ferromagnetic/metal/ferromagnetic).

Anisotropic Magnetoresistance

The physical origin of the AMR is given by the spin–orbit interaction and was first explained by Kondo [60] in 1962. The electrons that contribute to the electrical conduction in a metal such as Fe, Co or Ni are electrons in 4s and 3d states. The 4s electrons are considered free electrons and contribute largely to the conduction, while 3d electrons contribute to the conduction to only a certain extent because they are more localized. Empty or incomplete 3d states can be temporarily occupied by 4s conduction electrons. The characteristic property of transition metals is that the densities of spin-up and spin-down electrons in 3d states are different.

The 4s electrons are scattered depending on the direction of orbital angular momentum of the 3d electrons. Therefore, the resistance in the two channels is different. By applying an external magnetic field, the spins are oriented, then orbitals are reoriented, changing the electron scattering, and as a result changes in resistance are observed. In the case of polycrystalline cobalt or iron, a current applied parallel to the applied external field implies a higher resistance, owing to

Table 1.2 Comparison of different types of MR at room temperature

Type	Materials	MR (%)
AMR	Ferromagnetic	~5 %
GMR	Multilayer or granular (ferromagnetic/metal/ferromagnetic)	~20 %
TMR	Heterostructure or granular (ferromagnetic/insulator/ferromagnetic)	~200 %

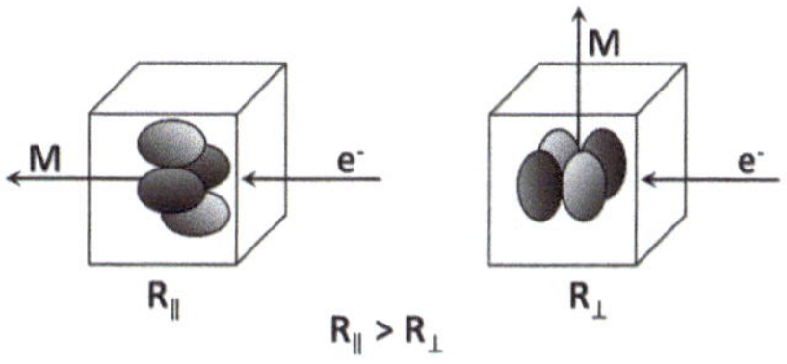

Fig. 1.14 The origin of AMR

the higher probability of an electron collision. A current applied perpendicular to the applied external field means a lower resistance, since the probability of an electron collision is smaller (see Fig. 1.14).

Tunnel Magnetoresistance in Granular Materials

The physical origin of TMR is given by the different probability of tunneling of electrons from a ferromagnetic material to another through an insulating barrier as a function of their spin orientation [61].

Granular systems are disordered composites formed by magnetic nanoparticles embedded in an insulating matrix. Thus, the electrical conduction mechanism may occur by tunneling between nanoparticles if the metal content is below the percolation threshold (20–60 at %). If nanoparticles are ferromagnetic, magnetization vectors will be misaligned in the absence of an external magnetic field, therefore in a state of high resistance. As we increase the magnetic field, the magnetization vectors are aligned, increasing the probability of tunneling of electrons between adjacent nanoparticles, going to a low resistance state (Fig. 1.15). There have been examples of this phenomenon in granular films of Co–Al–O, $FeSiO_2$ [62–64] and Fe-MgO [65].

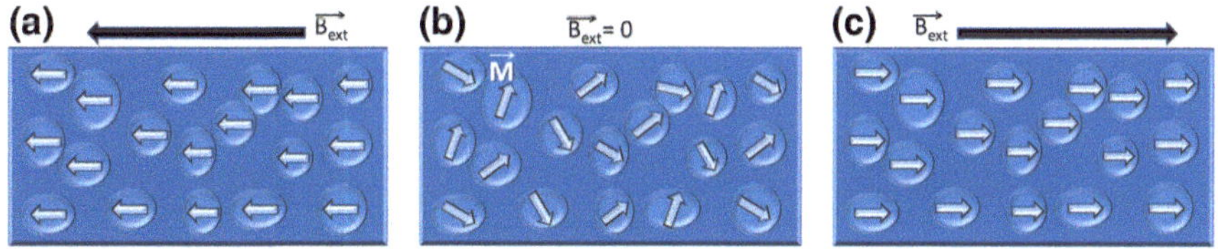

Fig. 1.15 Illustration of the magnetic configuration of the magnetic nanoparticles embedded in an insulating matrix. **a** The magnetization vectors are aligned with increasing applied magnetic field, increasing the probability that the conduction is carried out by the electron tunneling of nanoparticles, leading to a low resistivity state. **b** The magnetization vectors are misaligned in the absence of a magnetic field: high resistivity state. **c** As in (**a**), but increasing the applied magnetic field in the opposite direction

1.5.2.3 Hall Effect

The Hall effect is a voltage measured perpendicular to the current when a perpendicular magnetic field is applied. In the simplest case of a single type of carriers (electrons or holes), the Hall effect in non-magnetic materials is a linear function of applied magnetic field and its intensity is related to the density of charge carriers. The magnetization in ferromagnetic materials produces strong deviations from linearity, so the Hall resistivity (ρ_H) is defined by [60]

$$\rho_H = \frac{V_H t}{I} = \rho_{OHE} + \rho_{AHE} = \mu(R_0 H + R_S M) \tag{1.4}$$

where V_H is the measured Hall voltage, which is transverse to the current (I) direction and t is the film thickness. As mentioned, in magnetic materials, ρ_H is given by the ordinary contribution (ρ_{OHE}) related to the applied external magnetic field and the anomalous contribution (ρ_{AHE}) related to the magnetization, where R_0 and R_S are the ordinary and the anomalous Hall coefficients, respectively.

Anomalous Hall Effect

The anomalous Hall resistivity depends on the resistivity of the film as shown in Eq. (1.5) [66]:

$$\rho_{AHE} = a\rho + b\rho^2 \tag{1.5}$$

where a and b are constants. The linear term is given by the extrinsic scattering mechanism (skew scattering) of electrons in impurities and defects. The quadratic term is governed by the intrinsic contribution induced only by the band structure,

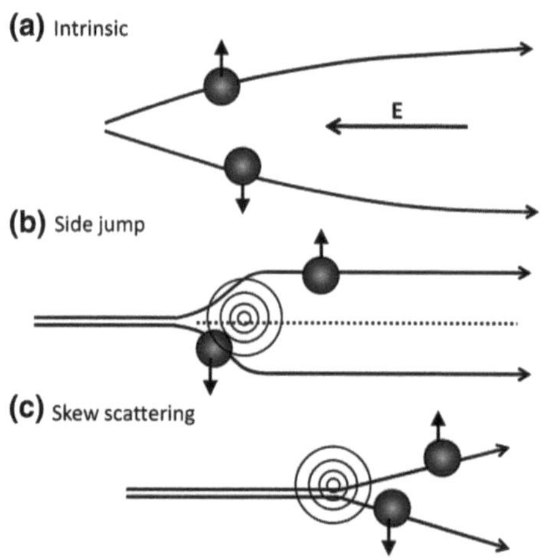

Fig. 1.16 Illustration of the three mechanisms that may govern the AHE. **a** Intrinsic contribution. **b** Side jump contribution. **c** Skew scattering contribution

which is independent of the electron scattering, or by the extrinsic scattering mechanism (side jump) of electrons [67–69]. The skew scattering describes the angle of deflection of the charge carrier's trajectory in the scattering. In the side jump the electron trajectory completely moves from its original path (Fig. 1.16).

1.6 Superconducting Nanostructures

1.6.1 Introduction to Superconductivity

Since Kamerling Onnes's discovery in 1911 of perfect conductivity (no resistance to the passage of electric current), which he called superconductivity [70], numerous related discoveries have been made, making it one of the main research areas in condensed matter physics [71]. Recent studies are related to superconductivity in one dimension [72], vortex confinement [73], vortex pinning [74], etc. Furthermore, we should not exclude the recent applications in the fabrication of nanoSQUIDS [75–77], superconducting quantum bits (SQBs) [78, 79], etc.

In 1986, Bednorz and Müller discovered a new class of high-temperature superconductors. Thus, the $LaBaCuO_4$ compound underwent a transition to the superconducting state at 30 K [80], and quickly the critical temperature (T_c) of this superconductor family reached 130 K. More recently, in 2008 Fe-based superconductors with $T_c \sim 50$ K were discovered [81]. These new findings lead us to new routes in the preparation and characterization of superconducting materials in order to implement them in the area of nanotechnology associated with condensed matter physics.

Going back to the beginnings of this new state, besides the lack of resistance below a critical temperature T_c, which is characteristic of each material, in 1933 Meissner and Ochsenfeld [82] discovered perfect diamagnetism. A superconducting material under a magnetic field ejects the field when is cooled below its T_c. This fact could be explained by the perfect conductivity phenomenon, which would tend to trap the magnetic flux. In fact, each material in the superconducting state is limited by the critical temperature, magnetic field and current density (see Fig. 1.17).

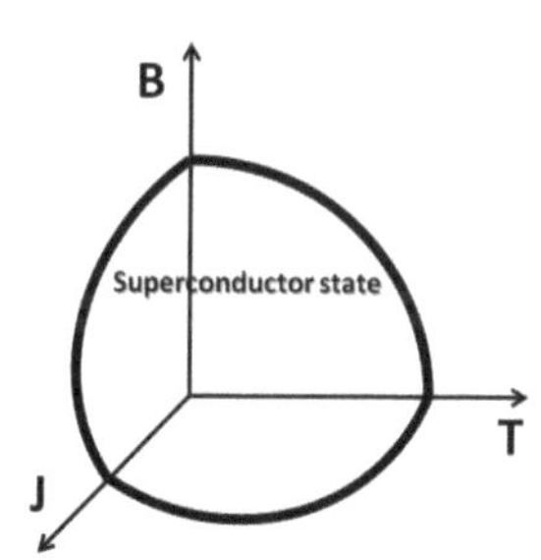

Fig. 1.17 Schematic superconducting state bounded by temperature, external magnetic field and current density, which are characteristic of each material

In a superconducting material the field is expelled by superficial supercurrents, which unscreen the external magnetic field. However, the field is not completely expelled at the surface, where supercurrents are generated, its value decreasing exponentially from the surface to a characteristic length, called the London penetration length, λ_L.

The macroscopic theory of Ginzburg and Landau (GL) [83] describes inhomogeneities or spatial variations of the superconducting electron density, n_s, introducing an order parameter $\Psi = |\Psi|e^{i\varphi}$ whose squared modulus $|\Psi|^2$ represents n_s. Thus, the GL theory can explain another superconductor material behavior regarding the external magnetic field different from the Meissner effect. When the magnetic field penetrates the superconductor, normal areas are generated that coexist with superconducting areas in which the field is expelled. The order parameter vanishes in these normal regions, while recovering its equilibrium position in superconducting regions.

The modulus of the order parameter completely vanishes in the normal region, recovering its initial value at a characteristic length, called the coherence length, ξ. As we mentioned, the magnetic field is not completely expelled from the superconducting material by the supercurrents produced, but it penetrates exponentially a characteristic length, from the surface to the normal core, called the penetration length, λ. The ratio of these two parameters, λ/ξ, is known as the Ginzburg–Landau parameter κ. If $\kappa > 1$, the field lines penetrate into the superconducting material, unlike systems that display the Meissner effect. Thus, one criterion is established to classify superconducting materials, based on their behavior under a magnetic field. Superconductors exhibiting the pure Meissner effect are called type I. However, most superconductors behave differently and are known as type II superconductors, in which the magnetic field is not completely unscreened in all regions of the material, and areas with normal behavior and superconducting regions, the mixed state, are distinguished.

In 1957, Abrikosov [84] explained how the magnetic field penetrates a type II superconductor. The field penetrates the superconductor by forming magnetic flux lines, called vortices. Each vortex has associated a flux quantum, $\phi_0 = h/2e = 2.07 \times 10^{-15}$ Wb. The vortex has a normal nucleus and circular supercurrents around it in the center, where the magnetic field is not completely ejected. The most favorable energetic configuration for the vortex arrangement owing to their interactions is a regular lattice with square or hexagonal symmetry.

The Bardeen–Cooper–Schrieffer (BCS) [85] microscopic theory explained the origin of the lack of resistance in the superconducting state. The idea is based on the weak attraction of two electrons close to the Fermi level, which are linked to form what is known as a Cooper pair with energy less than the Fermi energy E_F [86]. Such an attractive interaction is generated by the electron–phonon–electron interaction predicted in Ref. [87].

1.6.2 Vortex Dynamics

The vortices behave as objects that move in the presence of a current. The movement of the vortices is governed by [71]

$$\vec{f}_L - \eta \vec{\upsilon}_\varphi - \vec{f}_p = 0 \tag{1.6}$$

where the Lorentz force per unit length of the vortex is defined by $\vec{f}_L = \vec{J}_x \times \vec{\Phi}_0$, η is the viscosity of the vortex motion, $\vec{\upsilon}_\varphi$ is the vortex velocity and $\vec{f}_p$ is the pinning force.

When a current is applied, flow lines develop a Lorentz force per unit length, which tends to move vortices transverse to the applied current. This movement causes an energy dissipation, which generates an electric field and therefore induces a longitudinal voltage or finite resistance.

Following Faraday's law,

$$\vec{E} = -\vec{\nabla} U = \vec{B} \times \vec{\upsilon}_\varphi \tag{1.7}$$

The vortex motion induces the voltage $U_x = \Phi_0 \upsilon_{\Phi,y} / L$ at a given vortex density $n_\phi = B/\Phi_0$, so the number of vortices in a nanowire of length L and width w is $N = n_\phi L w$. The electric field is given by

$$E_x = N \frac{U_x}{w} = \frac{N \Phi_0 \upsilon_{\Phi,y}}{L w} B \upsilon_{\Phi,y} \tag{1.8}$$

Using Eq. (1.6), we find the electric field generated by the vortex motion is given by

$$E_x = \frac{B}{\eta} (f_L - f_p) \tag{1.9}$$

In most type II superconductors, the vortex motion is caused by the opposition of small inhomogeneities, which can pin vortices. Therefore a finite resistance is not observed until the Lorentz force exceeds the vortex pinning force. When the Lorentz force overcomes the vortex pinning force, the vortices move at a speed that depends on the material viscosity, and hence $E_x > 0$. When the pinning force is higher than the Lorentz force, vortices are pinned and $E_x = 0$.

In superconducting wires pinning is very strong, so that the current density that can be applied without energy dissipation is high. This will become clear in Chap. 5 of this thesis, where we present the electrical transport measurements in superconducting tungsten nanowires. The study of vortex pinning either occurring because of structural disorder or artificially induced by engineering patterns has been a major research line in superconductivity in recent decades [88].

1.6.2.1 Flux Flow Resistance

In the ideal case where there is no vortex pinning, the motion generated by the Lorentz force is prevented only by the force generated by the viscosity of the material. This is the pure flux flow regime. From expression (1.6), we obtain

$$J\Phi_0 = \eta v_\phi \tag{1.10}$$

$$\rho_f = \frac{E}{J} = B\frac{\Phi_0}{\eta c^2} \tag{1.11}$$

$$\eta = \frac{\Phi_0^2}{2\pi w^2 c^2 \rho_n} \approx \frac{\Phi_0 H_{c2}}{\rho_n c^2} \tag{1.12}$$

Substituting Eq. (1.12) in (1.11), we obtain a reduced expression for the flux flow resistivity:

$$\rho_f = \rho_n \frac{2\pi w^2 B}{\Phi_0} \approx \rho_n \frac{B}{B_{c2}} \tag{1.13}$$

Real materials have defects, caused by the presence of impurities, inhomogeneities, surface roughening, etc., which can generate pinning centers, since the vortices tend to occupy more energetically stable positions. This effect may be an advantage because it is a way to reduce energy dissipation by immobilizing flow lines, useful in high magnetic field solenoids.

In the flux creep phenomenon, thermal energy can allow flow lines to jump from one pinning center to another as a unit, in bundles, in response to the moving force generated by the current density. When no current is applied, as shown in Fig. 1.18(a), the probability of going from one energy minimum to the next to the left or right is the same whereas, with an applied current, Fig. 1.18(b), the vortex or vortices will move in the direction in which the current density decreases with

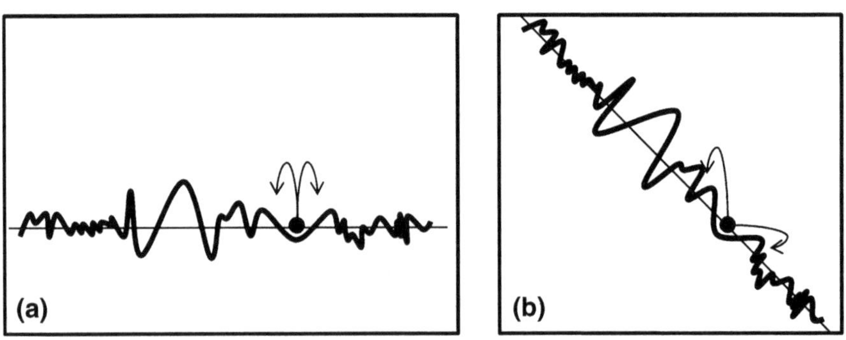

Fig. 1.18 Schematic representation of the vortex hopping energy barriers to adjacent pinning centers. **a** Motion force is zero. **b** Motion force due to the current favors jumps in the "downhill" direction

higher probability. Presumably, the jump rate of the flux lines would have an exponential dependence, as expressed in Eq. (1.14).

$$R \sim a\ e^{-\Delta F/kT} \tag{1.14}$$

The energy dissipation due to the movement of the vortices depends on the distribution of electrons in the core, which can produce nonlinear *V(I)* with very high current densities. There are two different cases of nonlinear effects: the Larkin and Ovchinnikov theory predicts flux flow instabilities at temperatures close to T_c, attributed to the vortex core shrinkage, and at temperatures much below T_c, where the effect is due to the heating of electrons [89].

1.6.2.2 Phase Slip Resistance

This section briefly discusses the mechanism of energy dissipation by phase slips, which can be seen in 1D superconductors, for example in nanowires with a thickness or width of the order of ξ [71, 72, 90–92] or structures with constrictions of dimensions comparable to ξ. The nanowire resistance is not zero below T_c owing to fluctuations of the superconducting order parameter when the dissipation mechanism is governed by phase slips. Thermal fluctuations are significant at temperatures close to T_c. The relative phase of the order parameter changes from one nanowire point to another by 2π, producing a finite voltage or resistance.

The theory of thermally activated phase slips (TAPS) was developed by Langer and Ambegaokar in 1967 [93] and completed by McCumber and Halperin in 1970 [94]. The LAMH theory explains the energy dissipation and finite resistance in 1D nanowires below T_c. The thermal activations of TAPS system occur at a rate governed by an exponential dependence as in

$$R \sim a\ e^{-\Delta F/kT} \tag{1.15}$$

when it exceeds a potential barrier ΔF.

There are different systems suitable for the study of vortex motion in superconducting materials. In Chap. 5, we will discuss the study of the vortex dynamics in tungsten microwires, nanowires, and ultranarrow nanowires grown by FIBID, a type II superconductor in the dirty limit, with weak pinning, which is ideal for studies of this type [95–97].

1.6.3 Nonlocal Vortices Dynamics

In 1965, Giaever [98] demonstrated experimentally flow line motion by means of electrical measurements and also the nonlocal motion of vortices in areas where the current is not flowing.

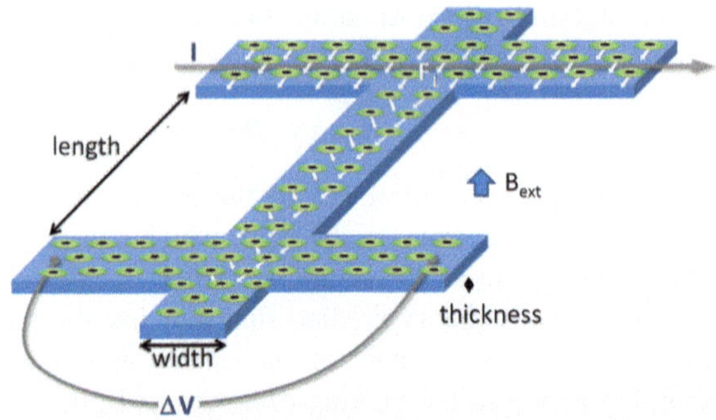

Fig. 1.19 Schematic diagram of the geometry of the sample used to measure the nonlocal voltage for samples with no current passing through the nanowire on the upper horizontal crossbar. The voltage drop generated by the movement of vortices was measured in the lower horizontal crossbar

The current flow in a type II superconductor structure and weak pinning as shown in Fig. 1.19 can cause a nonlocal voltage in a remote area that is connected by a channel where the vortices are moving.

The nonlocal voltage perpendicular to the direction of vortex motion was predicted theoretically by Wortis and Huse [99] and checked experimentally in high temperature superconductors, which have high viscosity and are close to the melting transition [100]. In principle this effect would disappear at distances greater than a few lattice parameters in the absence of circulating current [73].

Grigorieva et al. found a nonlocal effect, where the vortex motion in a type-II superconductor nanowire (NW) gave rise to a voltage measured in a transverse connection to a superconductor NW where there was no circulating current, such as shown in Fig. 1.19. The nonlocal motion of vortices remains over several hundreds of intervortex distances, in contrast to the dynamic effect induced by the viscosity [73, 101–104]. Thus, in type-II superconductors with weak pinning subjected to an applied magnetic field perpendicular to the plane, and an applied current in the upper horizontal ("short" in the figure) NW that acts on the vortices located in the upper cross region, pushing in the direction of the "long" NW (in the Fig. 1.19) due to the Lorentz force generates a nonlocal voltage in the lower cross region. The repulsion between vortices, in combination with the vortex density determined by the external magnetic field, leads to an efficient pressure transfer through the vortex lattice along the "long" NW, and once vortices are moved towards the lower cross region, a nonlocal voltage is generated corresponding to the electric field $\vec{E} = \vec{B} \times \vec{v}_{\phi}$.

In the framework of the study of the nonlocal vortex dynamics, tungsten FIBID nanowires were fabricated and characterized using the same configuration as given by Grigorieva. We will show the results in Chap. 5.

1.7 Objectives and Structure of This Thesis

The motivation of this thesis was the fabrication of functional materials at the nanoscale using novel techniques such as FEBID and FIBID. First, we will address the study of nanostructures that exhibit ferromagnetic behavior, such as the Co

FEBID in Chap. 3, and Fe FEBID in Chap. 4, whose most important applications are in the design of magnetic devices for the micro/nanoelectronics industry as memories, magnetic sensors, etc. Second, we will focus on the study of superconducting W nanostructures grown by FIBID, whose most notable applications are in the design of superconducting devices for the micro/nanoelectronics industry, such as making SQUIDs, non-finite resistance nanocontacts, and particle or photon sensors.

This thesis is grouped into six chapters and one appendix.

In Chap. 2 summarizes the main experimental techniques used throughout the thesis, both for the growth of nanostructures (SEM/FIB) and for characterization by means of microscopy techniques [SEM, atomic force microscopy (AFM), scanning transmission electron microscopy (STEM), high resolution transmission electron microscopy (HRTEM)], spectroscopy [energy dispersive X-ray spectroscopy (EDS), electron energy loss spectroscopy (EELS)], and magnetotransport measurements [physical properties measurement system (PPMS)].

The growth and characterization of ferromagnetic Co nanodeposits by FEBID are described in Chap. 3, showing how varying parameters has a major impact on their subsequent magnetic properties. Compositional analysis of nanodeposits was done in situ. The study of the microstructure and composition on the nanoscale was performed by transmission electron microscopy (TEM). The magnetic properties of ultranarrow nanostructures, 30 nm in lateral size, were studied by magnetotransport measurements.

Investigation of ferromagnetic Fe FEBID deposits is reported in Chap. 4. The tuning of Fe content and the microstructure allowed us to explore the metal-to-insulator transition, the evolution of the magnetoresistance, and the Hall effect.

In Chap. 5 we present the growth and characterization of superconducting W nanodeposits by FIBID. First, we study the influence of some growth parameters, such as beam acceleration voltage and the FIB incidence angle on the nanodeposits' composition. Second, we present the superconducting properties of W structures of different lateral sizes (from microwires to ultranarrow nanowires) obtained by local electrical transport measurements. Finally, we studied the generation of a nonlocal signal due to movement of the vortices by nonlocal electrical transport measurements.

The main conclusions and perspectives of this thesis are summarized in Chap. 6.

The appendix briefly describes the main characteristics of the Pt deposits grown by FIBID and FEBID.

The literature that was consulted during the work documented in this thesis is listed at the end of each chapter.

Finally, at the beginning of this thesis you will find a list of acronyms and a list of articles published as a result of this work.

References

1. Feynman, R.P.: There's plenty of room at the bottom. Eng. Sci. **23**, 22–36 (1960)
2. Spaldin, N.: In: Magnetic Materials Fundamentals and Device Applications. Cambridge University Press, Cambridge (2003)
3. Allwood, D.A., Xiong, G., Faulkner, C.C., Atkinson, D., Petit, D., Cowburn, R.P.: Magnetic domain-wall logic. Science **309**(5741), 1688–1692 (2005)
4. Novoselov, K.S., Geim, A.K., Morozov, S.V., Jiang, D., Zhang, Y., Dubonos, S.V., Grigorieva, I.V., Firsov, A.A.: Electric field effect in atomically thin carbon films. Science **306**(5696), 666–669 (2004)
5. Novoselov, K.S., Geim, A.K., Morozov, S.V., Jiang, D., Katsnelson, M.I., Grigorieva, I.V., Dubonos, S.V., Firsov, A.A.: Two-dimensional gas of massless Dirac fermions in graphene. Nature **438**(7065), 197–200 (2005)
6. Novoselov, K.S.: Nobel lecture: graphene: materials in the Flatland. Rev. Mod. Phys. **83**(3), 837–849 (2011)
7. Britnell, L., Gorbachev, R.V., Jalil, R., Belle, B.D., Schedin, F., Mishchenko, A., Georgiou, T., Katsnelson, M.I., Eaves, L., Morozov, S.V., Peres, N.M.R., Leist, J., Geim, A.K., Novoselov, K.S., Ponomarenko, L.A.: Field-effect tunneling transistor based on vertical graphene heterostructures. Science **335**(6071), 947–950 (2012)
8. van Kouwen, L., Botman, A., Hagen, C.W.: Focused electron-beam-induced deposition of 3 nm dots in a scanning electron microscope. Nano Lett. **9**(5), 2149–2152 (2009)
9. Jonckheere, R., Bret, T., Van den Heuvel, D., Magana, J., Gao, W., Waiblinger, M.: Repair of natural EUV reticle defects. In: Maurer, W., Abboud, F.E., (eds.) Photomask Technology 2011, vol. 8166. SPIE—The International Society of Optical Engineering, Bellingham (2011)
10. Giannuzzi, L.A., Stevie, F.A.: Introduction to Focused Ion Beams, p. 357. Springer Science, Boston (2005)
11. Reyntjens, S., Puers, R.: A review of focused ion beam applications in microsystem technology. J. Micromech. Microeng. **11**(4), 287–300 (2001)
12. Drouin, D.: Programa de simulación, CASINO (1996)
13. Drouin, D., Couture, A.R., Joly, D., Tastet, X., Aimez, V., Gauvin, R.: CASINO V2.42—a fast and easy-to-use modeling tool for scanning electron microscopy and microanalysis users. Scanning **29**(3), 92–101 (2007)
14. Gomer, R.: Mechanims of liquid-metal electron and ion sources. Appl. Phys. **19**(4), 365–375 (1979)
15. Ziegler, J.F.: Nuclear instruments & methods in physics research section B-Beam interactions with materials and atoms. **219**, 1027 (2004)
16. Rius, G., Llobet, J., Arcamone, J., Borrise, X., Perez-Murano, F.: Electron- and ion-beam lithography for the fabrication of nanomechanical devices integrated on CMOS circuits. Microelectron. Eng. **86**(4–6), 1046–1049 (2009)
17. Rius, G., Llobet, J., Borrise, X., Mestres, N., Retolaza, A., Merino, S., Perez-Murano, F.: Fabrication of complementary metal-oxide-semiconductor integrated nanomechanical devices by ion beam patterning. J. Vac. Sci. Technol., B **27**(6), 2691–2697 (2009)
18. Melngailis, J., Mondelli, A.A., Berry, I.L., Mohondro, R.: A review of ion projection lithography. J. Vac. Sci. Technol., B **16**(3), 927–957 (1998)
19. Tseng, A.A.: Recent developments in micromilling using focused ion beam technology. J. Micromech. Microeng. **14**(4), R15–R34 (2004)
20. Tseng, A.A.: Recent developments in nanofabrication using focused ion beams. Small **1**(10), 924–939 (2005)
21. Miller, M.K., Russell, K.F.: Atom probe specimen preparation with a dual beam SEM/FIB miller. Ultramicroscopy **107**(9), 761–766 (2007)
22. Miller, M.K., Russell, K.F., Thompson, G.B.: Strategies for fabricating atom probe specimens with a dual beam FIB. Ultramicroscopy **102**(4), 287–298 (2005)

23. Randolph, S.J., Fowlkes, J.D., Rack, P.D.: Focused, nanoscale electron-beam-induced deposition and etching. Crit. Rev. Solid State Mater. Sci. **31**(3), 55–89 (2006)
24. Martínez-Pérez, M.J., Sesé, J., Córdoba, R., Luis, F., Drung, D., Schuring, T.: Circuit edit of superconducting microcircuits. Supercond. Sci. Technol. **22**(12), 125020 (2009)
25. Randolph, S.J., Fowlkes, J.D., Rack, P.D.: Focused electron-beam-induced etching of silicon dioxide. J. Appl. Phys. **98**(3), 034902 (2005)
26. Yemini, M., Hadad, B., Liebes, Y., Goldner, A., Ashkenasy, N.: The controlled fabrication of nanopores by focused electron-beam-induced etching. Nanotechnology **20**(24), 245302 (2009)
27. Bret, T., Afra, B., Becker, R., Hofmann, T., Edinger, K., Liang, T., Hoffmann, P.: Gas assisted focused electron beam induced etching of alumina. J. Vac. Sci. Technol., B **27**(6), 2727–2731 (2009)
28. Wang, J.H., Griffis, D.P., Garcia, R., Russell, P.E.: Etching characteristics of chromium thin films by an electron beam induced surface reaction. Semicond. Sci. Technol. **18**(4), 199–205 (2003)
29. Wang, S., Sun, Y.M., White, J.M., Stivers, A., Liang, T.: Electron-beam-assisted etching of CrO_x films by Cl_2. J. Vac. Sci. Technol., B **23**(1), 206–209 (2005)
30. Ganczarczyk, A., Geller, M., Lorke, A.: XeF_2 gas-assisted focused-electron-beam-induced etching of GaAs with 30 nm resolution. Nanotechnology **22**(4), 045301 (2011)
31. Lassiter, M.G., Liang, T., Rack, P.D.: Inhibiting spontaneous etching of nanoscale electron beam induced etching features: solutions for nanoscale repair of extreme ultraviolet lithography masks. J. Vac. Sci. Technol., B **26**(3), 963–967 (2008)
32. Roediger, P., Wanzenboeck, H.D., Hochleitner, G., Bertagnolli, E.: Crystallinity-retaining removal of germanium by direct-write focused electron beam induced etching. J. Vac. Sci. Technol., B, **29**(4), 041801 (2011)
33. Schoenaker, F.J., Córdoba, R., Fernández-Pacheco, R., Magén, C., Stephan, O., Zuriaga-Monroy, C., Ibarra, M.R., De Teresa, J.M.: Focused electron beam induced etching of titanium with XeF_2. Nanotechnology **22**(26), 265304 (2011). doi:10.1088/0957-4484/22/26/265304
34. Santschi, C., Jenke, M., Hoffmann, P., Brugger, J.: Interdigitated 50 nm Ti electrode arrays fabricated using XeF_2 enhanced focused ion beam etching. Nanotechnology **17**(11), 2722–2729 (2006)
35. Matsui, S., Mori, K.: New selective deposition technology by electron-beam induced surface-reaction. Jpn. J. Appl. Phys. Part 2-Lett. **23**(9), L706–L708 (1984)
36. Matsui, S., Mori, K.: New selective deposition technology by electron-beam induced surface-reaction. J. Vac. Sci. Technol., B **4**(1), 299–304 (1986)
37. Koops, H.W.P., Weiel, R., Kern, D.P., Baum, T.H.: High resolution electron-beam induced deposition. J. Vac. Sci. Technol., B **6**(1), 477–481 (1988)
38. Silvis-Cividjian, N., Hagen, C.W.: Electron-beam-induced nanometer-scale deposition. In: Advances in Imaging and Electron Physics, vol 143. Elsevier Academic Press Inc: San Diego (2006)
39. van Dorp, W.F., Hagen, C.W.: A critical literature review of focused electron beam induced deposition. J. Appl. Phys. **104**(8), 081301 (2008)
40. Utke, I., Hoffmann, P., Melngailis, J.: Gas-assisted focused electron beam and ion beam processing and fabrication. J. Vac. Sci. Technol., B **26**(4), 1197–1276 (2008)
41. Wnuk, J.D., Gorham, J.M., Rosenberg, S.G., van Dorp, W.F., Madey, T.E., Hagen, C.W., Fairbrother, D.H.: Electron induced surface reactions of the organometallic precursor trimethyl(methylcyclopentadienyl)platinum(IV). J. Phys. Chem., C **113**(6), 2487–2496 (2009)
42. Fowlkes, J.D., Rack, P.D.: Fundamental electron-precursor-solid interactions derived from time-dependent electron-beam-induced deposition simulations and experiments. ACS Nano **4**(3), 1619–1629 (2010)
43. Silvis-Cividjian, N., Hagen, C.W., Kruit, P.: Spatial resolution limits in electron-beam-induced deposition. J. Appl. Phys. **98**(8), 084905 (2005)

44. Botman, A., Mulders, J.J.L., Hagen, C.W.: Creating pure nanostructures from electron-beam-induced deposition using purification techniques: a technology perspective. Nanotechnology **20**(37), 372001 (2009)
45. Córdoba, R., Sesé, J., De Teresa, J.M., Ibarra, M.R.: High-purity cobalt nanostructures grown by focused-electron-beam-induced deposition at low current. Microelectron. Eng. **87**(5–8), 1550–1553 (2010)
46. Mulders, J.J.L., Belova, L.M., Riazanova, A.: Electron beam induced deposition at elevated temperatures: compositional changes and purity improvement. Nanotechnology **22**(5), 055302 (2011)
47. Belova, L.M., Dahlberg, E.D., Riazanova, A., Mulders, J.J.L., Christophersen, C., Eckert, J.: Rapid electron beam assisted patterning of pure cobalt at elevated temperatures via seeded growth. Nanotechnology **22**(14), 145305 (2011)
48. Tripathi, S.K., Shukla, N., Rajput, N.S., Kulkarni, V.N.: The out of beam sight effects in focused ion beam processing. Nanotechnology **20**(27), 275301 (2009)
49. Hernández-Ramírez, F., Rodríguez, J., Casals, O., Russinyol, E., Vila, A., Romano-Rodríguez, A., Morante, J.R., Abid, M.: Characterization of metal-oxide nanosensors fabricated with focused ion beam (FIB). Sens. Actuators B-Chem. **118**(1–2), 198–203 (2006)
50. Li, W.X., Warburton, P.A.: Low-current focused-ion-beam induced deposition of three-dimensional tungsten nanoscale conductors. Nanotechnology **18**(48), 485305 (2007)
51. Hernández-Ramírez, F., et al.: Portable microsensors based on individual SnO_2 nanowires. Nanotechnology **18**(49), 495501 (2007)
52. Rius, G., Llobet, J., Esplandiu, M.J., Sole, L., Borrise, X., Perez-Murano, F.: Using electron and ion beams on carbon nanotube-based devices. Effects and considerations for nanofabrication. Microelectron. Eng. **86**(4–6), 892–894 (2009)
53. Marcano, N., Sangiao, S., Plaza, M., Perez, L., Fernández Pacheco, A., Córdoba, R., Sanchez, M.C., Morellón, L., Ibarra, M.R., De Teresa, J.M.: Weak-antilocalization signatures in the magnetotransport properties of individual electrodeposited Bi Nanowires. Appl. Phys. Lett. 96(8), 082110 (2010)
54. Muñoz-Rojas, D., Córdoba, R., Fernández-Pacheco, A., De Teresa, J.M., Sauthier, G., Fraxedas, J., Walton, R.I., Casan-Pastor, N.: High conductivity in hydrothermally grown $AgCuO_2$ single crystals verified using focused-ion-beam-deposited nanocontacts. Inorg. Chem. **49**(23), 10977–10983 (2010). doi:10.1021/ic101420c
55. Puers, R., Reyntjens, S.: Fabrication and testing of custom vacuum encapsulations deposited by focused ion beam direct-write CVD. Sens. Actuators A—Phys. **92**(1–3), 249–256 (2001)
56. Martínez-Pérez, M.J., Sesé, J., Luis, F., Córdoba, R., Drung, D., Schurig, T., Bellido, E., de Miguel, R., Gomez-Moreno, C., Lostao, A., Ruiz-Molina, D.: Ultrasensitive broad band SQUID microsusceptometer for magnetic measurements at very low temperatures. IEEE Trans. Appl. Supercond. **21**(3), 345–348 (2011)
57. Neugebauer, C.A., Webb, M.B.: Electrical conduction mechanism in ultrathin, evaporated metal films. J. Appl. Phys. **33**(1), 74–82 (1962)
58. Edwards, P.P., Ramakrishnan, T.V., Rao, C.N.R.: The metal-nonmetal transition—a global perspective. J. Phys. Chem. **99**(15), 5228–5239 (1995)
59. Baibich, M.N., Broto, J.M., Fert, A., Vandau, F.N., Petroff, F., Eitenne, P., Creuzet, G., Friederich, A., Chazelas, J.: Giant magnetoresistance of (001)Fe/(001)Cr magnetic superlattices. Phys. Rev. Lett. **61**(21), 2472–2475 (1988)
60. Kondo, J.: Anomalous Hall effect and magnetoresistance of ferromagnetic metals. Prog. Theoret. Phys. **27**(4), 772–792 (1962)
61. Julliere, M.: Tunneling between ferromagnetic-films. Phys. Lett. A **54**(3), 225–226 (1975)
62. Fujimori, H., Mitani, S., Ohnuma, S.: Tunnel-type GMR in Co-Al-O insulated granular system—its oxygen-concentration dependence. J. Magn. Magn. Mater. **156**(1–3), 311–314 (1996)
63. Mitani, S., Fujimori, H., Takanashi, K., Yakushiji, K., Ha, J.G., Takahashi, S., Maekawa, S., Ohnuma, S., Kobayashi, N., Masumoto, T., Ohnuma, M., Hono, K.: Tunnel-MR and spin

electronics in metal-nonmetal granular systems. J. Magn. Magn. Mater. **198–99**, 179–184 (1999)
64. Honda, S., Okada, T., Nawate, M., Tokumoto, M.: Tunneling giant magnetoresistance in heterogeneous Fe-SiO_2 granular films. Phys. Rev. B **56**(22), 14566–14573 (1997)
65. García-García, A., Vovk, A., Pardo, J.A., Strichovanec, P., Algarabel, P.A., Magén, C., De Teresa, J.M., Morellón, L., Ibarra, M.R.: Tunneling magnetoresistance in Fe/MgO granular multilayers. J. Appl. Phys. **107**(3), 033704 (2010)
66. Nagaosa, N., Sinova, J., Onoda, S., MacDonald, A.H., Ong, N.P.: Anomalous Hall effect. Rev. Mod. Phys. **82**(2), 1539–1592 (2010)
67. Smit, J.: The spontaneous Hall effect in ferromagnetics-I. Physica **21**(11), 877–887 (1955)
68. Smit, J.: The spontaneous Hall effect in ferromagnetics-II. Physica **24**(1), 39–51 (1958)
69. Berger, L.: Application of side-jump model to Hall-effect and Nernst effect in ferromagnets. Phys. Rev. B 1972, **5**(5) (1862)
70. Onnes, H. K.: Leiden Communications, vol. 120 (1911)
71. Tinkham, M.: Introduction to Superconductivity, 2nd edn. Dover Publications, Inc., New York (1996)
72. Arutyunov, K.Y., Golubev, D.S., Zaikin, A.D.: Superconductivity in one dimension. Phys. Rep.—Rev. Sect. Phys. Lett. **464**(1–2), 1–70 (2008)
73. Grigorieva, I.V., Geim, A.K., Dubonos, S.V., Novoselov, K.S., Vodolazov, D.Y., Peeters, F.M., Kes, P.H., Hesselberth, M.: Long-range nonlocal flow of vortices in narrow superconducting channels. Phys. Rev. Lett. **92**(23), 237001 (2004)
74. Velez, M., Martin, J.I., Villegas, J.E., Hoffmann, A., Gonzalez, E.M., Vicent, J.L., Schuller, I.K.: Superconducting vortex pinning with artificial magnetic nanostructures. J. Magn. Magn. Mater. **320**(21), 2547–2562 (2008)
75. Wu, C.H., Chou, Y.T., Kuo, W.C., Chen, J.H., Wang, L.M., Chen, J.C., Chen, K.L., Sou, U.C., Yang, H.C., Jeng, J.T.: Fabrication and characterization of high-T_c $YBa_2Cu_3O_{(7-x)}$ nanoSQUIDs made by focused ion beam milling. Nanotechnology **19**(31), 315304 (2008)
76. Foley, C.P., Hilgenkamp, H.: Why nanoSQUIDs are important: an introduction to the focus issue. Supercond. Sci. Technol. **22**(6), 064001 (2009)
77. Lam, S.K.H., Clem, J.R., Yang, W.: A nanoscale SQUID operating at high magnetic fields. Nanotechnology **22**(45), 455501 (2011)
78. Clarke, J., Wilhelm, F.K.: Superconducting quantum bits. Nature **453**(7198), 1031–1042 (2008)
79. Leggett, A.J.: Physics—Superconducting qubits—a major roadblock dissolved? Science **296**(5569), 861–862 (2002)
80. Bednorz, J.G., Müller, K.A.: Possible high-Tc superconductivity in the Ba-La-Cu-O system. Zeitschrift für Physik B—Condens. Matter **64**(2), 189–193 (1986)
81. Kamihara, Y., Watanabe, T., Hirano, M., Hosono, H.: Iron-based layered superconductor $La[O_{1-x}F_x]FeAs$ (x = 0.05-0.12) with T_c = 26 K. J. Am. Chem. Soc. **130**(11), 3296 (2008)
82. Meissner, W., Ochsenfeld, R.: Short initial announcements. Naturwissenschaften **21**, 787–788 (1933)
83. Ginzburg, V.L., Landau, L.D.: On the theory of superconductivity. Zh. Eksperim. i Teor. Fiz. **20**, 1064 (1950)
84. Abrikosov, A.A.: On the magnetic properties of superconductors of the second group. Sov. Phys. JETP-USSR **5**(6), 1174–1183 (1957)
85. Bardeen, J., Cooper, L.N., Schrieffer, J.R.: Theory of superconductivity. Phys. Rev. **108**(5), 1175–1204 (1957)
86. Cooper, L.N.: Bound electron pairs in a degenerate Fermi gas. Phys. Rev. **104**(4), 1189–1190 (1956)
87. Fröhlich, H.: Theory of the superconducting state.1. The ground state at the absolute zero of temperature. Phys. Rev. **79**(5), 845–856 (1950)
88. Blatter, G., Feigel'man, M.V., Geshkenbein, V.B., Larkin, A.I., Vinokur, V.M.: Vortices in high-temperature superconductors. Rev. Modern Phys., **66**(4), 1125–1388 (1994)

89. Larkin, A.I., Ovchinnikov, Y.N.: Vortex motion in superconductors. In: Nonequilibrium Superconductivity. North-Holland, Amsterdam (1986)
90. Bezryadin, A.: Quantum suppression of superconductivity in nanowires. J. Phys.—Condens. Matter, **20**(4), 043202 (2008)
91. Li, P., Wu, P.M., Bomze, Y., Borzenets, I.V., Finkelstein, G., Chang, A.M.: Switching currents limited by single phase slips in one-dimensional superconducting Al nanowires. Phys. Rev. Lett. **107**(13), 137004 (2011)
92. Papari, G., Carillo, F., Stornaiuolo, D., Longobardi, L., Beltram, F., Tafuri, F.: High critical current density and scaling of phase-slip processes in YBaCuO nanowires. Supercond. Sci. Technol. **25**(3), 035011 (2012)
93. Langer, J.S., Ambegaokar, V.: Intrinsic resistive transition in narrow superconducting channels. Phys. Rev. **164**(2), 498 (1967)
94. McCumber, D.E., Halperin, B.I.: Time scale of intrinsic resistive fluctuations in thin superconducting wires. Phys. Rev. B—Solid State, **1**(3), 1054 (1970)
95. Sadki, E.S., Ooi, S., Hirata, K.: Focused-ion-beam-induced deposition of superconducting nanowires. Appl. Phys. Lett. **85**(25), 6206–6208 (2004)
96. Guillamón, I., Suderow, H., Vieira, S., Fernández-Pacheco, A., Sesé, J., Córdoba, R., De Teresa, J.M., Ibarra, M.R.: Nanoscale superconducting properties of amorphous W-based deposits grown with a focused-ion-beam. New J. Phys. **10**(9), 093005 (2008)
97. Guillamón, I. Orden y desorden en superconductividad. Thesis, Universidad Autónoma de Madrid (2009)
98. Giaever, I.: Magnetic coupling between 2 adjacent type-2 superconductors. Phys. Rev. Lett. **15**(21), 825 (1965)
99. Wortis, R., Huse, D.A.: Nonlocal conductivity in the vortex-liquid regime of a two-dimensional superconductor. Phys. Rev. B **54**(17), 12413–12420 (1996)
100. López, D., Kwok, W.K., Safar, H., Olsson, R.J., Petrean, A.M., Paulius, L., Crabtree, G.W.: Spatially resolved dynamic correlation in the vortex state of high temperature superconductors. Phys. Rev. Lett. **82**(6), 1277–1280 (1999)
101. Helzel, A., Kokanovic, I., Babic, D., Litvin, L.V., Rohlfing, F., Otto, F., Surgers, C., Strunk, C.: Nonlocal vortex motion in mesoscopic amorphous $Nb_{0.7}Ge_{0.3}$ structures. Phys. Rev. B **74**(22), 220510 (2006)
102. Otto, F.: Nonlinear vortex transport in mesoscopic channel of amorphous NbGe. Thesis, Universitätsverlag Regensburg (2009)
103. Otto, F., Bilusic, A., Babic, D., Vodolazov, D.Y., Suergers, C., Strunk, C.: Nonlocal versus local vortex dynamics in the transversal flux transformer effect. Phys. Rev. B **81**(17), 174521 (2010)
104. Otto, F., Bilusic, A., Babic, D., Vodolazov, D.Y., Suergers, C., Strunk, C.: Reversal of nonlocal vortex motion in the regime of strong nonequilibrium. Phys. Rev. Lett. **104**(2), 027005 (2010)

Chapter 2
Experimental Techniques

In this chapter we present the experimental techniques used in the development of this thesis, for the fabrication of samples and for subsequent characterization. The sample fabrication was carried out in three "Dual Beam" (SEM/FIB) systems with similar characteristics: Nova 200 NanoLab and Helios NanoLab 600 stationed in the Advanced Microscopy Laboratory, Institute of Nanoscience of Aragon (INA-LMA), University of Zaragoza, and the Nova 600 NanoLab installed at the Eindhoven University of Technology (TU/e). The characterization of the samples was performed in situ (SEM, EDS, electrical transport measurements) and ex situ (AFM, PPMS, HRTEM, STEM/EELS,...) in the Advanced Microscopy Laboratory, Institute of Nanoscience of Aragon (LMA-INA) and the Physical Measurements Service of the University of Zaragoza.

2.1 Nanolithography by "Dual Beam"

2.1.1 Description of a "Dual Beam"

In Chap. 1 we provided an overview of a "Dual Beam" system, detailing the fundamental parts that make it up and its wide range of applications. Owing to the important role that this tool has played in the development of our research, including the fabrication of samples, we definitely consider it "the heart" of this thesis. Figure 2.1 shows a photograph of the Helios NanoLab 600 instrument, one of the "Dual Beams" used for fabrication of nanostructures.

In Fig. 2.2 we can see inside the process chamber of the Nova 200 NanoLab (left; another "Dual Beam" widely used for the growth of nanostructures in the development of this thesis) and the Helios NanoLab 600 (right). Given the importance of these systems in the development of this thesis, we now indicate which aspects of the equipment or tools have been used specifically.

R. Córdoba Castillo, *Functional Nanostructures Fabricated by Focused Electron/Ion Beam Induced Deposition*, Springer Theses, DOI: 10.1007/978-3-319-02081-5_2,

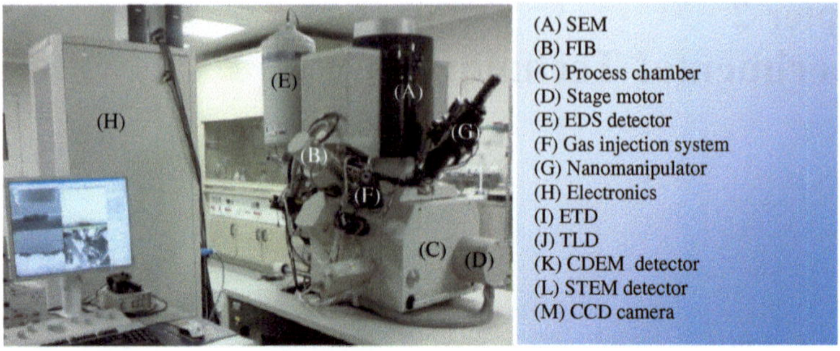

Fig. 2.1 Photograph of the Helios NanoLab 600 "Dual Beam" system installed in the clean room of the LMA-INA. (*ETD* Everhart–Thornley detector, *TLD* through-the-lens detector, *CDEM* continuous-dynode electron multiplier)

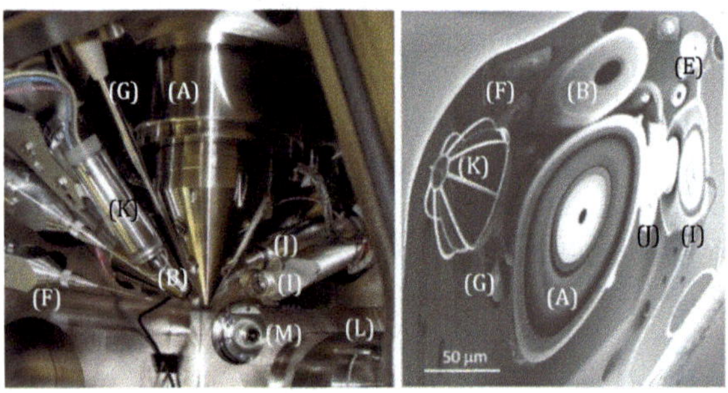

Fig. 2.2 Inside the process chamber of two Dual Beams. *Left* Photograph of the Nova 200 NanoLab installed at the LMA-INA. *Right* SEM image. An insulating material to be charged by the FEB irradiation acts as an "electronic mirror" and allows us to get a fascinating picture of some pieces of the Helios NanoLab 600 from the perspective of the sample

1. Acquisition of high resolution images using the appropiate SEM, STEM and FIB detectors
2. Milling by FIB: cross sections, etc.
3. Nanofabrication by SEM: growth of Co, Fe and Pt nanodeposits
4. Nanofabrication by FIB: growth of W and Pt nanodeposits
5. Compositional analysis of nanodeposits by the EDS technique
6. Electrical characterization of the fabricated nanodeposits by the four-probe technique
7. Manipulation of micro and nanostructures using a nanomanipulator
8. TEM sample preparation.

2.1.2 SEM

The electron column is based on a field emission electron source, with a W filament. The optics of the column focuses the electron beam to obtain the best resolution: 0.9 nm operating at 15 kV, 11 pA.

2.1.3 FIB

The ion column can focus the ion beam down to 5 nm in diameter using the smallest aperture of the system (~ 1 pA), which gives us the ability to mill or grow ultranarrow structures on the order of a few nanometers. The maximum aperture of the Magnum column in the Nova 200 NanoLab and Helios NanoLab 600 allows us to operate with a current of ~21 nA (beam diameter ~411 nm), very useful for milling large areas or trenches, as in TEM sample preparation.

2.1.4 Gas Injection System and Precursor Materials

2.1.4.1 Gas Injection System

The gas injection system (GIS) is compact and individual for each precursor material, as seen in Fig. 2.3. Both insertion and retraction of the GIS and the opening of the valve that allows the gas precursor to enter the process chamber are very fast.

The temperature of the crucible, where the precursor material is stored, and the needle, which is screwed to it, is homogeneous and controlled. Therefore, the pressure in the crucible is constant and, as a consequence, the flow of molecules that is released through the needle is only related to the saturation vapor pressure

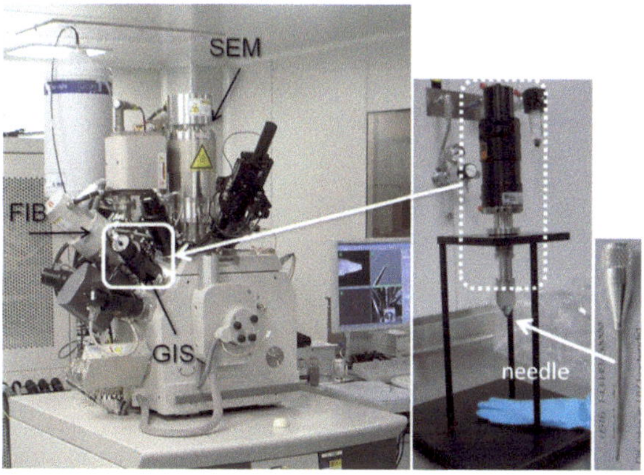

Fig. 2.3 Photograph of a "Dual Beam" in which the port where the GIS is installed is indicated. In the smaller photographs we see the needle through which the precursor gas leaves (*right*) and the GIS on a platform (*center*)

and the resistance to the gas flow caused by the needle. The vapor pressure of a precursor material is given by the Clausius–Clapeyron equation, where $\log(P) \propto 1/T$. Given this equation, a change of ±0.1 °C of the crucible temperature would result in a variation in pressure of about 1 %. The GIS allows the temperature to be controlled within that range of values, so that the flow delivered by the needle may have a variation of ±1 %. The valve is opened automatically in the GIS, except in the case of Co and Fe precursors, when it is opened manually to avoid an abrupt rise in the chamber pressure. Moreover, the precursor temperature should not be too high to prevent the precursor chemistry being modified by natural thermal decomposition [1]. The temperatures of the precursor materials used in the development of this thesis were between 22 and 55 °C.

In addition to the gas flow exiting the GIS needle, it is important to know the sample area that is covered by precursor gas molecules. This area depends on several factors, such as the incidence angle of the needle, the separation of the needle from the substrate surface, the distance from the needle tip to the area of interest on the sample, and the shadow effect [2].

2.1.4.2 Precursor Materials

Most of the precursor materials used in FEBID and FIBID are organometallics that are widely known and implemented in other growth techniques such as CVD. For a simple implementation and installation in the "Dual Beam" is preferable that they are solid or liquid at room temperature (with exceptions such as xenon difluoride,

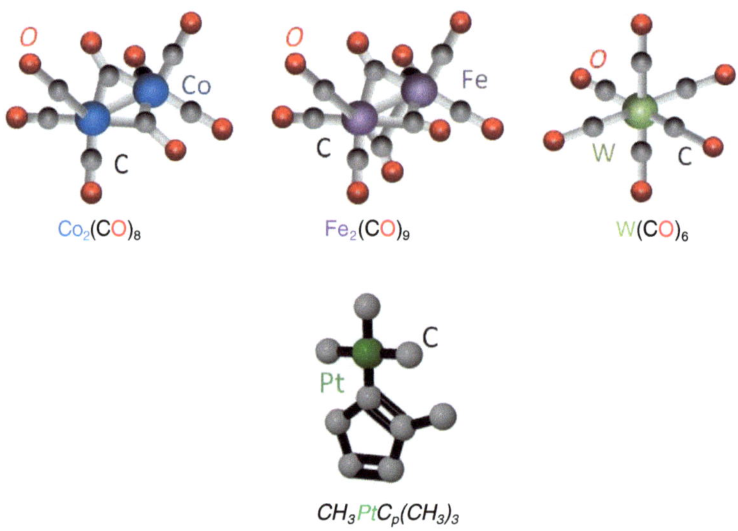

Fig. 2.4 Precursor materials used in this work for nanostructure growth

XeF_2) and their sublimation temperature is close to room temperature. Other considerations to take into account in the selection of precursor materials suitable for FEBID and FIBID are based on toxicity and health risks.

The following precursor materials were used in this work (see Fig. 2.4):

1. Dicobalt octacarbonyl, $Co_2(CO)_8$, GIS temperature = 27 °C
2. Diiron nonacarbonyl, $Fe_2(CO)_9$, GIS temperature = 28 °C
3. Tungsten hexacarbonyl, $W(CO)_6$, GIS temperature = 55 °C
4. Methylcyclopentadienyl (trimethyl) platinum, $(CH_3)_3(CH_3C_pPt)$, GIS temperature = 45 °C

The different nature of each precursor material influences the decomposition under the FEB or FIB. Therefore, the nanostructures grown with each precursor exhibit different composition, microstructure and physical properties as detailed in the following chapters.

2.2 Micro-Hotplate

In this section we briefly explain the experimental setup we used to study the influence of substrate temperature on the decomposition of the precursor material $Co_2(CO)_8$ during nanostructure growth. The results are given in Sect. 3.4 (Chap. 3).

The main element of the experimental setup is a micro-hotplate based on a Pt layer, which uses the Joule effect to convert electrical power into heating power to introduce a current by a DC voltage or current source. Thus, the temperature could be increased up to ~650 °C.

For electrical access from outside the microscope to micro-hotplate, we designed and fabricated some connection pieces, which are essential elements of the experimental setup. We list the most important elements here:

1. Commercial micro-hotplate KMHP-100 (Kebaili Corporation) based on a Pt film (~400 nm thick) embedded in a suspended membrane type of sandwich of SiO_2 (~500 nm thick)/Si_3N_4 (~1.5 μm thick)/SiO_2 (~400 nm thick), with a working area of about 250 × 150 μm (see Fig. 2.5d).
2. Support plate with four terminals to fit the micro-hotplate to the four metal contacts which are bonded to the four wires (see Fig. 2.5c).
3. Sample holder in which is installed the support plate mentioned previously, indicated by a green rectangle in Fig. 2.5a, which is fully engaged in the process chamber (see Fig. 2.5b).
4. Connector to the port of the door of the process chamber, indicated by a white rectangle in Fig. 2.5a.
5. Connector from the door of the process chamber to a connection box, marked by a blue square in Fig. 2.5a.
6. DC voltage source.
7. Connection box.

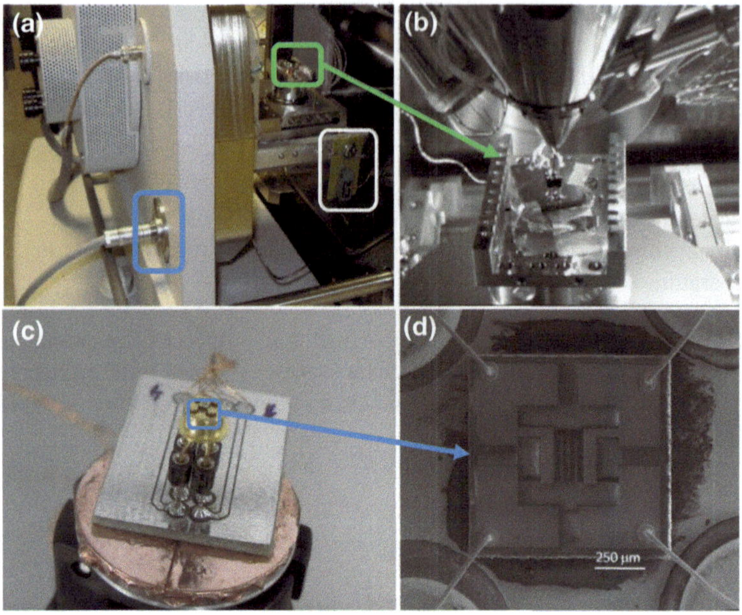

Fig. 2.5 **a** Photograph of the open door of the process chamber of the "Dual Beam". The *blue rectangle* on the *left* indicates the connection from the computer. On the *right* we can see the platform where the micro-heater is installed (indicated by the *green rectangle*). The *white rectangle* marks the connection to the port in the door of the process chamber. **b** Photograph of the inside of the process chamber of the "Dual Beam", where we can observe the micro-hotplate. **c** Photograph of the micro-hotplate inserted in the sample holder, which is installed in the process chamber. **d** SEM image of the micro-hotplate connected to four terminals

Fig. 2.6 Schematic representation of the physical basis of the EDS technique

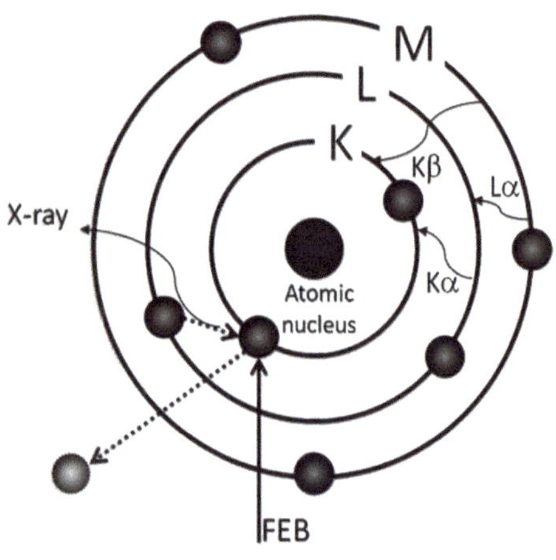

2.3 Energy Dispersive X-Ray Spectroscopy

Compositional analysis of nanostructures grown by FIBID and FEBID was carried out in situ inside the "Dual Beam" by the energy dispersive X-ray spectroscopy detector (Oxford Instruments) (see Figs. 2.1 and 2.2), which collects the X-ray signal generated when the FEB scans the sample.

The incident electrons produce excited atomic states and the photon energy of X-rays emitted in the de-excitation is specific to each element, which allows semi-quantitative determination of the sample composition using the ZAF [3] method (see Fig. 2.6). This technique is simple to use and particularly sensitive to heavier elements.

The FEB interaction volume depends primarily on FEB acceleration voltage, FEB incidence angle with the substrate and the nature of the substrate. The energy resolution is on the order of ~125 eV and the spatial resolution depends on the interaction volume and FEB probe used (see Fig. 2.7). Thus, in the analysis of nanostructures as in our case we take special care in selecting the parameters to obtain the most representative spectrum of the nanostructure fabricated. As nanodeposits from different precursors have a completely different composition, we must select the best conditions for collection of individual spectra.

Just to find a starting point for all of them, if we analyze nanowires with lateral size and thickness below 100 nm, we should use a low FEB accelerating voltage to decrease the interaction volume, and a probe or FEB diameter <100 nm whenever we want to analyze elements that have atomic transitions below the voltage chosen. Thus, we try to avoid atomic transitions of the substrate in the spectrum, which we should eliminate in the data processing.

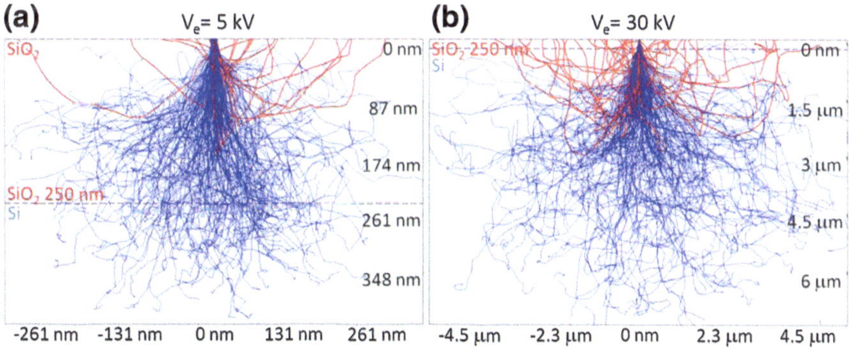

Fig. 2.7 Simulation of the FEB interaction volume on a thermally oxidized silicon substrate by the program CASINO (*"monte CArlo SImulation of electroN trajectory in sOlids"*) [4] at (**a**) 5 kV and (**b**) 30 kV

2.4 Structural and Compositional Characterization at the Nanoscale

2.4.1 TEM Sample Preparation by "Dual Beam"

The recent technological advances in transmission electron microscopy, which allow us to study individual atoms, mean that for these studies an appropriate sample preparation has become the cornerstone of a successful subsequent analysis. The thickness of these samples should be less than 100 nm, allowing the FEB to pass through the sample and the transmitted electrons to be collected. For the acquisition of high-resolution images samples should be even thinner, $\sim$50 nm thick.

The continued development of techniques based on the use of a "Dual Beam" has reduced the time for and complexity of the TEM sample preparation process. As example of this is the use of a specific holder for the whole process of the lamellae preparation for a cross-sectional view without breaking the vacuum in the chamber. In Fig. 2.8, we show a photograph of this specimen holder installed on the stage of the microscope, indicating the sample from which we will take one slice and the standard Cu grid used as a support, where the lamella will be welded for further manipulation and TEM study.

In the preparation process, we combine several techniques available on the "Dual Beam", such as deposition by FEBID and FIBID, milling by FIB, and nanomanipulator control. Moreover, as we work on an area of only about 50 μm^2, the rest of the sample is preserved for other types of studies.

Next, we detail the following steps in the lamella preparation [5–8]:

1. Pt FEBID. The Pt layer protects the area of interest during sputtering by the FIB (see Fig. 2.9a). Thus Pt FEBID prevents the damage or amorphization that can occur in the first nanometers of the surface of a Pt layer grown by FIBID. Deposit dimensions: 15 $\times$ 2 $\times$ 0.1 μm (length $\times$ width $\times$ thickness).

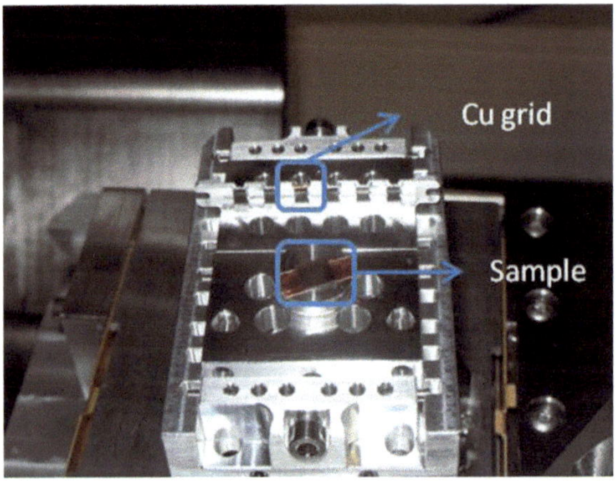

Fig. 2.8 Photograph of a specific sample holder for TEM sample preparation

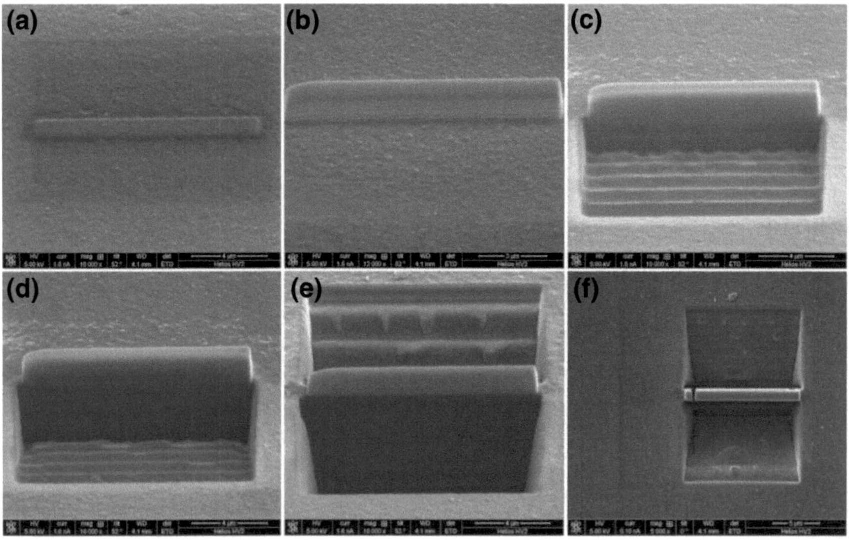

Fig. 2.9 SEM images illustrating stages of the TEM sample preparation, excluding the lift-out and final thinning and polishing

2. Pt FIBID. A film is deposited on top of that previously grown by FEBID to protect the area of interest during the following stages of milling (see Fig. 2.9b). Deposit dimensions: 15 × 2 × 1 μm (length × width × thickness).
3. Rough milling by FIB. Material on both sides of the area covered with Pt is removed by FIB (see Fig. 2.9c–e). Trench dimensions: 18 × 9 × 4 μm (length × width × depth). Lamella thickness after this stage is ~1 μm.
4. Fine milling by FIB. Material on both sides of the area covered with Pt is removed by FIB. Trench dimensions: 18 × 1 × 4 μm (length × width × depth). Lamella thickness after this stage is ~0.5 μm.
5. FIB cutting with an inverted U-shape. This does not completely release the lamella from the substrate (see Fig. 2.9f).
6. A film is deposited on that previously grown by FEBID to protect the area of interest during the following stages of grinding (see Fig. 2.9b).
7. Lift-out by an Omniprobe nanomanipulator, model 200. The nanomanipulator is a sharp tungsten tip of 1 μm diameter. It is inserted into the process chamber and its movements are motorized in the plane and on the z-axis with nanometer precision (maximum resolution of 10 nm). Thus, the nanomanipulator approaches the interest area until almost touching the lamella (see Fig. 2.10a). Then, we “weld” the nanomanipulator to the lamella by a Pt FIBID deposit and release the lamella from the substrate by three FIB cuts, with inverted U-shape (see Fig. 2.10b). Once the lamella has been released from the substrate and attached to the nanomanipulator (see Fig. 2.10c), we move the nanomanipulator towards the TEM grid (see Figs. 2.10c) and (d)) and grow

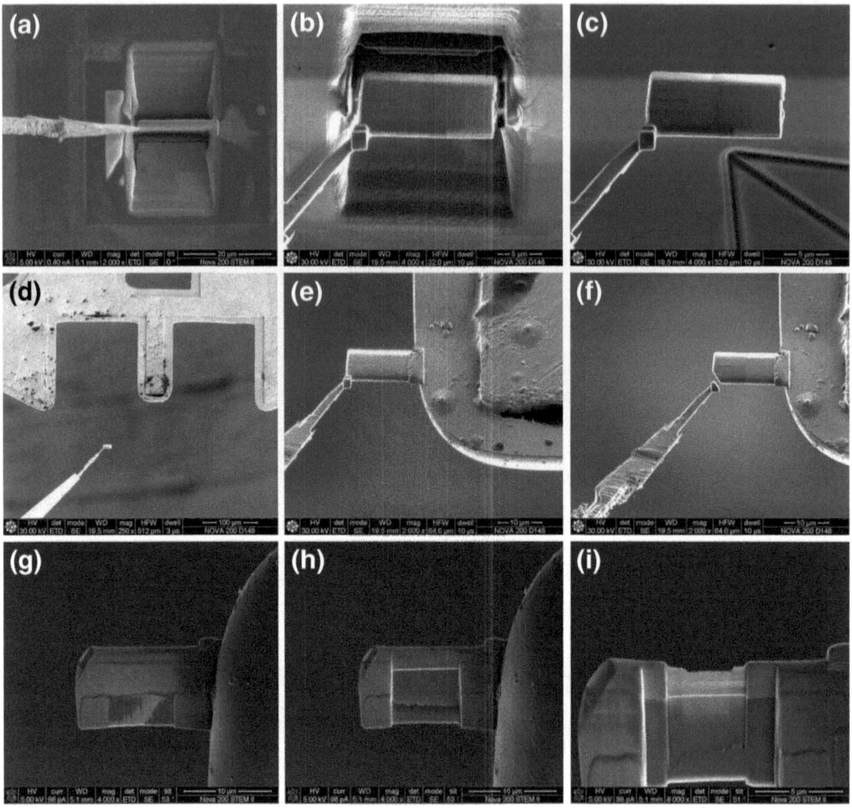

Fig. 2.10 SEM and FIB images illustrating the most delicate stages of the TEM sample preparation: the lift-out and the thinning and polishing

the Pt FIBID to weld the other end of the lamella to the grid (see Fig. 2.10e). Finally, we release the nanomanipulator from the grid by a FIB cut (see Fig. 2.10f). After the lift-out, we sharpen the nanomanipulator by FIB for the next use and retract it to its initial position.

8. Thinning. Fine milling by FIB. Material on both sides of the area covered with Pt is removed by FIB. Trench dimensions: 10 $\times$ 0.1 $\times$ 2 μm (length $\times$ width $\times$ depth) (see Fig. 2.10g). The process is repeated while the design length is reduced by about ~4 μm until the lamella thickness is ~50 nm (see Fig. 2.10h).
9. Polishing by FIB at low voltage (<5 kV). Cut dimensions: 4 $\times$ 0.01 $\times$ 1 μm (length $\times$ width $\times$ depth). The process is repeated until the thickness of the lamella is ~30 nm (see Fig. 2.10i).
10. Transfer of the Cu grid and the welded lamella to a microscope (TEM/STEM) for imaging and analysis study.

Polishing is carried out at low FIB voltage because the thickness of the amorphization layer generated in the lamella depends on this voltage. It can reach around 20 nm below its surface when the FIB operates at 30 kV. However, if we reduce the voltage to 2 kV, the thickness of the damaged layer is reduced to ~2 nm.

The complete process of TEM cross-sectional sample preparation can take several hours, but it is still much faster and less destructive than manual sample preparation, which requires a large number of processes, such as successive cuts of the sample using a diamond saw, gluing the sample in the sandwich shape, curing of the glue in a furnace, mechanical polishing, ion polishing, … This can take from several hours to a few days of work depending on the type of material being worked. Some examples of the results obtained thanks to excellent FIB lamella preparation at LMA-INA are reflected in the following references [9–15].

2.4.2 Transmission Electron Microscopy

Transmission electron microscopy (TEM) is a technique to obtain structural information from a very thin sample (thickness <100 nm). The electron beam is relatively wide and simultaneously illuminates the entire working region through the sample thickness; its interaction is reflected in the transmitted electrons, from which the projected image is generated in real space and the corresponding diffraction diagram in reciprocal space. Normally, the beam accelerating voltage is from 100 to 300 kV. The TEM image can be generated from diffracted or scattered electrons using a variety of contrast mechanisms that provide information about the morphology, crystallography and composition of the sample, with a spatial resolution at the atomic scale of 0.1 nm. Do not forget that this provides a two-dimensional projection of three-dimensional objects.

The HRTEM images presented in Chaps. 3 and 4 were taken at the Advanced Microscopy Laboratory, Institute of Nanoscience of Aragon, University of Zaragoza, by Dr. César Magén using a Titan [3] High-base (FEI) at 300 kV with spherical aberration corrector, C_s (spherical aberration coefficient). The spherical aberration correction of the objective lens can achieve a spatial resolution of at least 0.1 nm.

The HRTEM images of Pt deposits grown by FEBID and FIBID summarized in Appendix A and W deposits grown by FIBID summarized in Chap. 5 were taken at the Scientific-Technical Services of the University Barcelona by Dr. Jordi Arbiol, with a JEOL 2010F at 200 kV, 0.19 nm spatial resolution.

2.4.3 *Scanning Transmission Electron Microscopy*

A transmission electron microscope can be modified or converted to a scanning transmission electron microscope by adding a beam scanning system. Beam deflection is performed by scanning coils. In fact, you can carry out STEM using a SEM column at relatively low energies (30 kV) or, as mentioned before, by a high-voltage TEM column (100–300 kV). High voltages provide better resolution and the electrons are able to pass through thicker samples. The high-energy STEM resolutions achieved are close to those of TEM.

2.4.4 *Electron Energy Loss Spectroscopy*

In electron energy loss spectroscopy (EELS), the material to be tested is exposed to an electron beam known kinetic energy distribution in a small range. Some of the primary electrons undergo inelastic scattering, lose energy—which is transferred to the electrons or sample atoms—and are deviated randomly from their path. The amount of energy lost by the electrons can be measured by an electron spectrometer and interpreted in terms of the source of that energy loss. The main types of mechanisms of inelastic interactions include phonon excitation, plasmon excitation, electron excitation, and incident radiation loss.

This technique measures the composition, the chemical bond and the valence at the nanoscale. It usually works better for light elements of low atomic number and the energy resolution can be of the order of ~ 1 eV.

The fine structure analysis is performed by the ELNES (energy-loss near-edge structure) determination method, which gives details of the ionization structure and thus of the bond associated with each atomic species. The white lines are well-defined spectra, with very abrupt transitions due to narrow d or f bands.

The STEM images and EELS spectra in Chap. 3 of this thesis were taken and analyzed by Dr. Rodrigo Fernández-Pacheco in the STEM Group, Laboratoire de Physique des Solides (CNRS-UMR 8502), Université Paris-Sud, using a STEM VG HB 501 at 100 kV and a Gatan 666 spectrometer, optically coupled to a CCD camera.

2.5 Ex Situ Electric Transport Characterization

The electrical characterization of our samples in a wide temperature range from 0.3 to 300 K and applied magnetic field perpendicular to the substrate from +9 to –9 T was performed using a commercial system, the “Physical Properties Measurements System” (PPMS), from Quantum Design at the Department of Scientific Instrumentation, Physical Measurement Area, University of Zaragoza

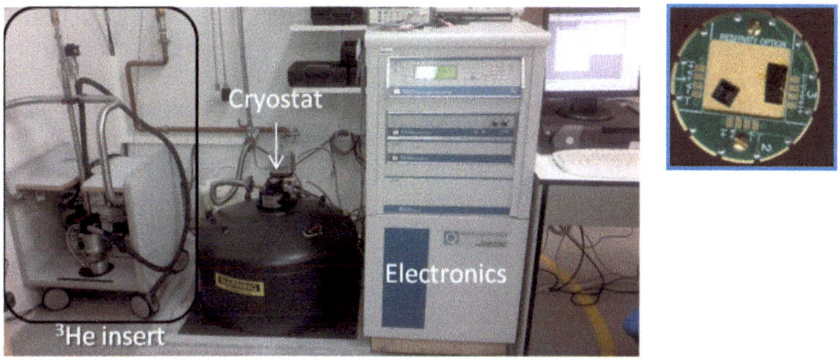

Fig. 2.11 Photograph of PPMS equipment (Quantum Design) installed in the Physical Measurements Service at the University of Zaragoza. In the inset, we show a photograph of the specimen holder provided with three independent measurement channels, each with four points, and used in the resistivity option

(see Fig. 2.11). In most cases, the option used for the characterization of our FIBID and FEBID deposits by four contacts was the resistivity in AC mode. In specific cases, as in Chap. 5, the resistivity in DC and AC mode with a ^{3}He cooling system was used. For measurements below 2 K, the PPMS system has the option to work with a ^{3}He refrigeration system that cools to 300 mK (see Fig. 2.11).

References

1. FEI www.fei.com
2. Friedli, V., Utke, I.: Optimized molecule supply from nozzle-based gas injection systems for focused electron- and ion-beam induced deposition and etching: simulation and experiment. J. Phys. D. Appl. Phys. **42**(12), 125305 (2009)
3. Castaing, R., Descamps, J.: Sur les bases physiques de l'analyse ponctuelle par spectrographie-X. J.phys. et le Radium. **16**(4), 304–317 (1955)
4. Drouin, D., Couture, A.R., Joly, D., Tastet, X., Aimez, V., Gauvin, R.: CASINO V2.42–A fast and easy-to-use modeling tool for scanning electron microscopy and microanalysis users. Scanning **29**(3), 92–101 (2007)
5. Giannuzzi, L.A., Stevie, F.A.: Introduction to focused ion beams, p. 357. Springer Science, Boston (2005)
6. Langford, R.M., Rogers, M.: In situ lift-out: Steps to improve yield and a comparison with other FIB TEM sample preparation techniques. Micron **39**(8), 1325–1330 (2008)
7. Mayer, J., Giannuzzi, L.A., Kamino, T., Michael, J.: TEM sample preparation and FIB-induced damage. MRS Bull. **32**(5), 400–407 (2007)
8. Young, R.J., Kirk, E.C.G., Williams, D. A., Ahmed, H.: Fabrication of planar and cross-sectional TEM specimens using a focused ion-beam, vol. 199, pp. 205–216. Materials Research Society, Pittsburgh (1990)
9. De Teresa, J.M., Córdoba, R., Fernández-Pacheco, A., Montero, O., Strichovanec, P., Ibarra, M.R.: Origin of the difference in the resistivity of as-grown focused-ion- and focused-electron-beam-induced Pt nanodeposits. Journal of Nanomaterials **2009**, 936863 (2009)

10. Sangiao, S., Morellon, L., Simon, G., Teresa, J.M.D., Pardo, J.A., Arbiol, J., Ibarra, M.R.: Anomalous hall effect in Fe (001) epitaxial thin films over a wide range in conductivity. Phys. Rev. B (Condensed Matter Mater. Phys.) **79**(1), 014431 (2009)
11. Estradè, S., Rebled, J.M., Arbiol, J., Peiró, F., Infante, I.C., Herranz, G., Sanchez, F., Fontcuberta, J., Córdoba, R., Mendis, B.G., Bleloch, A.L.: Effects of thickness on the cation segregation in epitaxial (001) and (110) $La_{2/3}Ca_{1/3}MnO_3$ thin films. Appl. Phys. Lett. **95**(7), 072507 (2009)
12. Marcano, N., Sangiao, S., Magen, C., Morellon, L., Ibarra, M.R., Plaza, M., Perez, L., De Teresa, J.M.: Role of the surface states in the magnetotransport properties of ultrathin bismuth films. Phys. Rev. B **82**(12), 125326 (2010)
13. Estradè, S., Rebled, J.M., Walls, M.G., de la Pena, F., Colliex, C., Córdoba, R., Infante, I.C., Herranz, G., Sanchez, F., Fontcuberta, J., Peiro, F.: Effect of the capping on the local Mn oxidation state in buried (001) and (110) $SrTiO_3/La_{2/3}Ca_{1/3}MnO_3$ interfaces. J. Appl. Phys. **110**(10), 103903 (2011)
14. Lavrijsen, R., Córdoba, R., Schoenaker, F.J., Ellis, T.H., Barcones, B., Kohlhepp, J.T., Swagten, H. J.M., Koopmans, B., De Teresa, J.M., Magén, C., Ibarra, M.R., Trompenaars, P., Mulders, J.J.L.: Fe:O:C grown by focused-electron-beam-induced deposition: magnetic and electric properties. Nanotechnology. **22**(2), 025302 (2011)
15. Córdoba, R., Fernández-Pacheco, R., Fernández-Pacheco, A., Gloter, A., Magén, C., Stephan, O., Ibarra, M.R., De Teresa, J.M.: Nanoscale chemical and structural study of Co-based FEBID structures by STEM-EELS and HRTEM. Nanoscale Res. Lett. **6**, 1–6 (2011)

Chapter 3
Ferromagnetic Cobalt Nanostructures Grown by Focused Electron Beam Induced Deposition

In this chapter we present the fabrication and characterization of Co nanodeposits, showing that varying the growth parameters such as the electron current, the beam accelerating voltage, the precursor flow, and the substrate temperature has significant impact on subsequent magnetic properties. Compositional analysis of these nanodeposits was done in situ.

The study of the microstructure and composition at the nanoscale was performed using transmission electron microscopy techniques, whereas the magnetic properties of ultranarrow nanostructures were studied by magnetotransport measurements. The results obtained allow us to link Co nanostructures with potential applications in the field of nanotechnology.

3.1 Introduction

The interest in ferromagnetic FEBID nanostructures arises from their promising applications owing to their magnetic character. The intrinsic properties of the deposits allow them to be used in a variety of applications in nanotechnology, sensor fabrication [1], magnetic data storage [2], magnetic logic [3], and biosensors [4].

In the growth of magnetic nanostructures by FEBID, the concentration of the magnetic element in the deposit is crucial owing to the degradation of the magnetic properties in the presence of residual material of the precursor, such as C and O. One of the most successful deposits fabricated and characterized by FEBID is the one based on cobalt using dicobalt octacarbonyl, $Co_2(CO)_8$, as a precursor material [5–9]. These Co deposits consist of cobalt nanocrystals 3–5 nm in size embedded in an amorphous carbon matrix.

In the literature, different strategies are given for improving the metal content of Co deposits grown by FEBID, for example, by modifying various deposition parameters such as the electron current [6], or using in situ or ex situ heat treatment [10].

Recently, some researchers have begun using cobalt tricarbonyl nitrosyl $Co(CO)_3(NO)$ [11, 12] as the precursor material for Co-based structures grown by

R. Córdoba Castillo, *Functional Nanostructures Fabricated by Focused Electron/Ion Beam Induced Deposition*, Springer Theses, DOI: 10.1007/978-3-319-02081-5_3,

FEBID. However, their results and the exhibited magnetic properties are not better than those obtained using $Co_2(CO)_8$ as the precursor material.

In this chapter, we describe the detailed investigation of Co deposits grown by FEBID using $Co_2(CO)_8$ as the precursor material. We pay special attention to the deposition parameters; they are crucial to obtain ferromagnetic nanostructures 30–100 nm in lateral size. So, the control of deposition parameters is essential to obtain high-Co-content nanostructures. We carried out a study of fabrication of Co nanostructures, modifying some of the parameters, such as FEB acceleration voltage (V_e), the electron current (I_e), the substrate temperature, the substrate nature, and the precursor flow. These nanostructures were characterized in situ or ex situ using advanced microscopy and spectroscopy techniques such as EDS, STEM, HRTEM, EELS, electronic holography, and magnetotransport measurements.

3.2 Growth of Cobalt Nanostructures by FEBID

The fabrication of Co nanostructures was carried out by focused electron beam induced deposition (FEBID), using a gas injection system (GIS) in a Nova Nanolab 200 "Dual Beam". This growth process was explained in detail in Sect. 1.3.4 (Chap. 1). The deposition parameters of Co nanodeposits are the precursor material (dicobalt octacarbonyl, $Co_2(CO)_8$); precursor material temperature (27 °C ± 1 °C); volume per dose (VD ~ 5×10^{-4} μm³/nC); substrate temperature ($T_{substrate}$ = room temperature), dwell time (DT = 1 μs), overlap (OL = 50 %); refresh time (RT = 0s); base pressure in the process chamber ($P_{base} = 1 \times 10^{-6}$ mbar; process pressure ($P_{process} = 6 \times 10^{-6}$ mbar; scanning strategy (serpentine mode, bottom to top); range of acceleration voltage (5 kV $\leq V_e \leq$ 30 kV); electron current range (21 pA $\leq I_e \leq$ 9.5 nA. Particular cases in which we modify any of these parameters will be discussed in the text. The substrates used in the experiments were boron-doped silicon (p-type doping) that had been thermally oxidized (with this method we obtain a 250 nm layer of insulating material (SiO_2)) and Si_3N_4 membranes of 50 nm thickness. Ti contacts were made only on the oxidized Si substrate using optical lithography techniques.

The first Co nanodeposits studied in this section had dimensions 0.5 μm × 1 μm × 0.2 μm and were grown on a thermally oxidized Si substrate with varying I_e and V_e to study the effect of these two parameters on the nanodeposits' composition.

The compositional analysis of Co nanodeposits was performed in situ by EDS. Thus, the deposits were not exposed to environmental conditions, avoiding degradation processes by external agents. The spectra were collected using an acceleration voltage of the electron beam of 5 kV. In Table 3.1, we show the results obtained in different Co nanostructures grown with varying I_e and V_e; concentrations above 90 % are achieved when I_e is in the range of nanoamps.

Table 3.1 Composition of Co nanodeposits grown on a thermally oxidized Si substrate to study the dependence of Co content on I_e and V_e

Sample	V_e (kV)	I_e (nA)	Co (at.% ± 2 %)	C (at.% ± 2 %)	O (at.% ± 2 %)
1	5	0.40	83	15	2
2	5	1.60	83	13	4
3	5	6.30	91	7	1
4	10	0.13	81	9	10
5	10	0.54	92	5	3
6	10	2.10	95	4	1
7	10	8.40	94	5	1
8	18	9.20	88	9	2
9	30	0.15	62	19	19
10	30	9.50	97	3	–

In Fig. 3.1, we show as an example the composition dependence of nanodeposits grown at 10 kV as a function of I_e, where we can clearly see that the Co content increases from 80 % at $I_e = 0.13$ nA to 95 % above 1 nA.

Utke and colleagues previously found this dependence [6]. However, the maximum Co content obtained did not exceed ~80 %, probably because their equipment included a thermionic tungsten filament and therefore the current density of electrons generated is considerably lower than that achieved by a field emission gun as in our system. Another reason would be that the base pressure in our system is one order of magnitude lower than in their microscope; the chamber contamination thus contributes significantly to the lower concentration of Co obtained. The accepted explanation of this phenomenon in most publications is that the heat generated locally by the electron currents in the nanoamp range of the electron beam emitted by a field emission gun facilitates the decomposition of this precursor material, whose decomposition temperature is relatively low compared with other precursors of a similar nature [10, 13].

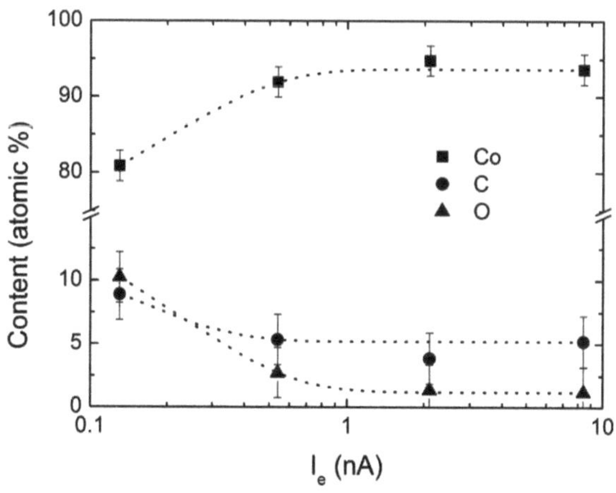

Fig. 3.1 Composition of Co nanodeposits grown at 10 kV as a function of I_e

3.3 Microstructure and Composition of Cobalt Nanodeposits at the Nanoscale

After noting the differences observed in the Co content between FEBID deposits grown by low and high electron current, we continue our study of the microstructure and composition at the nanoscale. The aim of this study is to analyze the valence state and the crystal structure of Co deposits. We want to explain the chemical properties and structural and magnetic properties of the nanodeposits studied by our group [13] from the chemical and structural points of view at the micro- and nanoscales.

The most appropriate technique to perform this type of local observation is (S)TEM. Specifically, EELS is the ideal technique to analyze the oxidation state and the chemical environment at the local scale of the three elements present in the deposit: carbon, oxygen, and cobalt. In STEM, EELS spectra can be determined and related to their position in the sample by the simultaneous acquisition of high angle annular dark field (HAADF) images. Moreover, the acquisition and analysis of HRTEM images can be used to extract information about the crystalline structure at the atomic scale.

The high metal content in the nanostructures grown using electron currents in the studied nanoamp range was confirmed by both techniques. Co nanodeposits were fabricated (at 30 kV on a thermally oxidized Si wafer, 250 nm SiO_2) using various values of I_e to study the influence of the electron beam current in the microstructure and composition of the Co-based nanodeposits at the nanoscale. The other parameters were the same as those used in Sect. 3.2.

Deposit 1 was grown at low I_e (44 pA) and deposit 2 at high I_e (2.4 nA) with dimensions 0.5 μm × 1.0 μm × 0.2 μm (width × length × thickness), pitch (p_{FEB}) ~2.21 nm (for 44 pA), and ~13.16 nm (for 2.4 nA).

After growth of the nanodeposits, EDS analysis was performed in situ. Results [at.%] for deposits 1 and 2 are 64:17:19 (Co:C:O) and 93:5:2 (Co:C:O), respectively. Before the preparation of TEM samples or lamellae in transverse view, the Co deposits were covered with two protective layers, first a layer of Pt by FEBID and then a second layer of Pt by FIBID. This standard procedure discussed in detail in Chap. 2 is performed to protect the Co deposits from the damage (amorphization) that may be caused by the FIB during the lamellae preparation. The preparation and extraction process of the lamellae of Co deposits is performed in situ in the same "Dual Beam", without breaking the vacuum in the process chamber. Thinning and final polishing of the lamellae is performed using an ion beam acceleration voltage (V_i) of 5 kV, to decrease the thickness of the amorphization layer produced by FIB. Finally, we obtain high quality lamellae with a thickness of ~30 nm for subsequent studies in which we intend to achieve atomic resolution.

As mentioned previously, the microstructure of the nanodeposits was investigated by HRTEM, performed on a Titan [3] high-base (FEI) with C_s spherical aberration corrector at 300 kV. The spherical aberration correction of the objective lens allows us to obtain a spatial resolution below 0.1 nm. The composition of nanodeposits at the nanoscale was investigated by chemical analysis performed using a STEM VG HB 501 with a cold field emission gun (cold-FEG) operated at 100 kV, and a Gatan 666 spectrometer optically coupled to a CCD camera.

For each sample, we conducted a chemical and structural analysis at the nanoscale by EELS and HRTEM. Together with a chemical analysis of the inside of each deposit, we also performed analyses of Pt–Co and SiO_2–Co interfaces to understand the differences in chemical composition at the center and the deposit surface. Spatially resolved EELS analysis was used to investigate the metallic cobalt content and the oxidation state in each deposit. Thus, the electron beam was scanned over the sample, and a series of spectra were collected for each point; thus, the obtained spectra can be compared as a function of the point of collection in the sample. This technique is known as spectrum-line or line scan acquisition [14]. For each spectrum-line, spectra were acquired at steps of 1 nm, and then summed every five spectra for the calculation of intensity ratios of the Co $L_{2,3}$ edge ($I(L_2)$ and $I(L_3)$, respectively). $I(L_2)$ and $I(L_3)$ were calculated as the intensity maximum for each edge and then compared with reference values $I(L_2)/I(L_3) = 0.31$ and 0.27 for Co metal and Co oxide (CoO), respectively [15].

For the analysis of chemical composition as a function of growth direction, 200 spectra were acquired for each point, realigned, and summed. Principal component analysis (PCA) was applied to each series of spectra to decrease experimental noise and thus obtain a better signal-to-noise ratio [16]. After PCA had been applied to each spectrum for a single point, five resulting consecutive spectra of a spectrum-line were summed, and the intensities of the white lines were calculated after a power-law removal of the background and a linear fit below the lines. Therefore, the chemical state of Co was first estimated by means of the intensity ratio of the L_2 and L_3 peaks.

Additionally, the relative O/Co concentrations were also calculated, integrating their respective signal intensities from a series of 200 summed EELS spectra at a single point inside the deposit and dividing by their respective cross sections. An energy dispersion of 0.2 eV/channel was used for the analysis of the fine structure for each element, whereas an energy dispersion of 0.5 eV/channel was used for the quantification of the relative amounts of each element, with a collection angle of 24 mrad and a convergence angle of 7.5 mrad. Both types of experiments had an acquisition time of 0.8 s/spectrum.

Direct observation of the HRTEM images shows that the inside of the deposit is made of polycrystalline cobalt nanoparticles (ca. 2–3 nm nanocrystal size) embedded in an amorphous carbon matrix (Fig. 3.2). The presence of such small nanoparticles had been previously reported in the literature [7]. The HRTEM image is dominated by the amorphous contrast of the matrix, which gives rise to a fast Fourier transform (FFT) blurred by diffuse scattering. Thus, only weak reflections associated with metallic hexagonal close packed (hcp) Co can be identified. Precise quantitative analysis of these kinds of granular samples is not

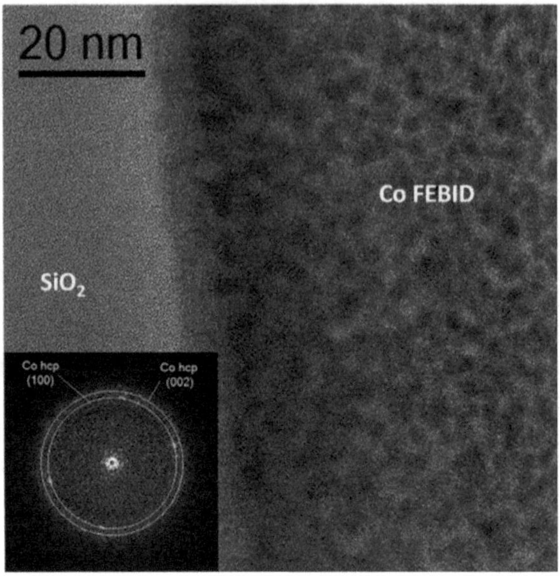

Fig. 3.2 HRTEM image and FFT (inset) of deposit 1. Figure reprinted from [17]

feasible. The presence of metallic cobalt and cobalt oxide species is evident from the in situ compositional EDS analysis, where 19 % (atomic) O content is observed.

On the other hand, to identify the oxidation state of Co, the study of the $L_{2,3}$ edge of cobalt and the K edge of oxygen in the EELS spectra can be very useful. The obtained spectra can be compared to EELS data in the literature to check for a shift in energy or any variation in the shape of the edges.

Figure 3.3a shows the O K edge of deposit 1 collected at different points of the sample. Thus, we confirm the existence of oxygen already in the spectrum collected at the core of the deposit, as observed by EDS. Furthermore, the presence of a small pre-peak at 531 eV in the O K edge fine structure of the deposit and the interface (not observed in the SiO_2 spectrum) is a distinctive sign of the presence of CoO [18]. Also, the ELNES analysis of the Co $L_{2,3}$ edge can yield very useful information. Thus, the L_2/L_3 intensity ratio between the peaks of the white lines of the cobalt spectrum gives us an indication of the oxidation state of Co: when L_2/L_3 decreases the oxidation state increases [19].

Figure 3.4 is a comparison of the white lines of the Co $L_{2,3}$ edge for deposits 1 and 2, and references of metallic Co and CoO. The EELS analysis for the first deposit shows the presence of oxidized cobalt, as can be inferred from the shape of the $L_{2,3}$ edge of the cobalt spectrum and the low average L_2/L_3 ratio of $\sim$0.27.

Deposit 2 has a different microstructure and composition from deposit 1. In the HRTEM image shown in Fig. 3.5, Co nanocrystals of 7–10 nm are presented. Co grains are more evenly distributed and compacted within the deposit. The crystalline structure obtained by indexing the digital diffractogram is compatible with a mixture of hcp and fcc Co (see FFT in Fig. 3.5).

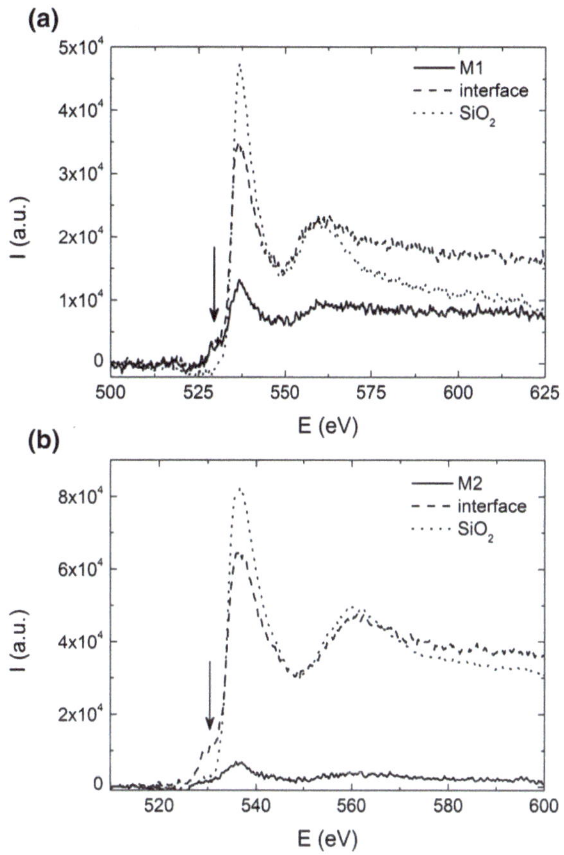

Fig. 3.3 The O K edge spectra from the SiO_2/Co interface were collected for deposits 1 **a** and 2 **b**. As the probe scans across the SiO_2 substrate, the interface between the two materials, and finally the inner part of the deposit, the O K edge changes its shape (appearance of a small pre-peak, marked by an arrow), practically disappearing at the inside of the microstructure for deposit 2. Figure adapted from [17]

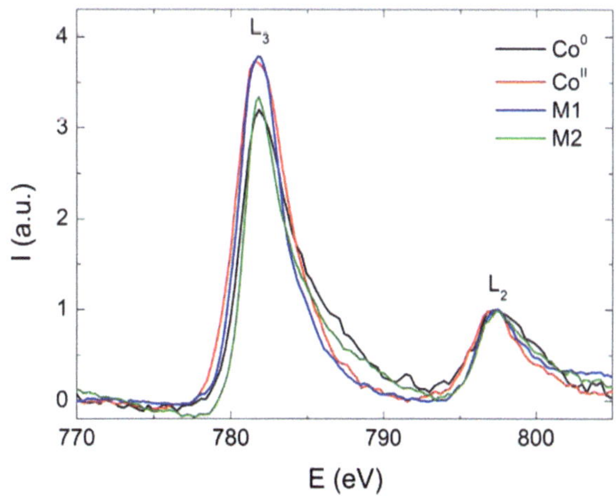

Fig. 3.4 Comparison of the EELS spectra of the Co $L_{2,3}$ edge (at an energy of 779 eV) for deposits 1 and 2, and metallic cobalt and cobalt(II) oxide as references. Figure reprinted from [17]

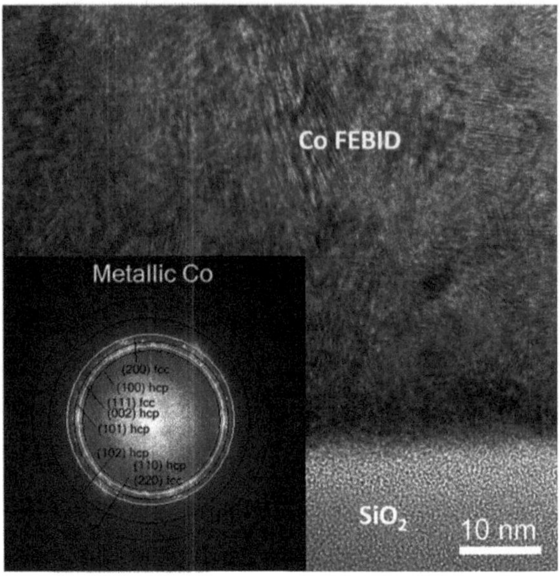

Fig. 3.5 HRTEM image and FFT (inset) of deposit 2. Figure adapted from [17]

Regarding the EELS spectra, the ELNES study of the cobalt $L_{2,3}$ edge yielded homogeneous and regular spectra with the characteristic white lines of metallic cobalt (Fig. 3.4). Indeed, for metallic Co, the L_3 line shows a broad asymmetric shape compared to the narrower L_3 line of Co oxide. Therefore, the metallic character of the nanocrystals is confirmed by the L_2/L_3 ratio of 0.30 and negligible oxygen content (O/Co ratio ~0.04).

On the other hand, Fig. 3.3b shows the EELS spectra at the O K edge region at the SiO_2/Co interface. Looking into the fine structure at the interface between the SiO_2 substrate and the cobalt structure, for the first nanometers of the growth of the deposit, one can observe the presence of a pre-peak at 531 eV, which is characteristic of the presence of CoO. As the probe scans the inner part of the deposit, the oxygen signal practically disappears. The presence of the CoO could be due to the existence of water molecules adsorbed on the substrate before the start of the FEBID process.

Table 3.2 is a summary of the preparation conditions for both samples and the quantitative ratio between oxygen and cobalt inside the deposit. The ELNES analysis yields information about the shape and the intensity of the major features

Table 3.2 Summary of deposition parameters, beam voltage (V_e) and current (I_e), EELS quantification ratio between oxygen and cobalt, and the average L_2/L_3 intensity ratios in the Co $L_{2,3}$ edge. Table reprinted from [17]

Deposit	V_e (kV)	I_e (nA)	O/Co	$I(L_2)/I(L_3)$
1	30	0.044	0.85	0.27
2	30	2.400	0.04	0.30

for both the Co $L_{2,3}$ and O K edges. In order to estimate the oxidation state of cobalt, the intensity ratio between the peaks L_2/L_3 of the Co $L_{2,3}$ edges was analyzed. As expected from previous EDS analyses, the deposit grown at a high beam current (nanoamp range) presents a lower O/Co ratio and a higher L_2/L_3 intensity ratio (close to that of metallic cobalt) than that grown at a low beam current. Therefore, EELS analysis shows that deposit 2 presented features characteristic of metallic cobalt, a fact confirmed by the absence of the O K edge for this particular deposit. On the other hand, oxidized cobalt was found in deposit 1, as can be inferred from the shape of the $L_{2,3}$ edge of the cobalt spectrum and the high L_2/L_3 ratio, as well as from the presence of a characteristic pre-peak at 531 eV for the O K edge feature.

However, HRTEM images for deposit 1 revealed the presence of hcp Co, a fact confirmed by the EELS analysis, which showed minor features of metallic cobalt. To understand the presence of CoO together with metallic Co in samples grown at a low beam current, we can assume that the particles that build up the deposit are so small that most of their atoms are present on the surface, oxidizing very easily and in a large proportion.

Figure 3.6 illustrates the homogeneity in composition and metallicity along the direction of deposition for deposit 2. A relative quantification of the elements has been performed as a function of the growth direction of the deposit, confirming the metallic state of cobalt. The ratio O/Co is very low, lower than 0.1 all along the deposit. Only the first nanometers of deposition seem to be partially oxidized. This is in good agreement with the plotting of the $I(L_2)/I(L_3)$ intensity ratios along the deposit, which shows metallic ratios throughout the deposit except in the early stages of growth, where the intensity ratio falls to 0.27 (Fig. 3.6b).

Summarizing, in the growth conditions which are the same with those used in Sect. 3.2 and in our previous publications [9, 13, 20], electron beam current plays a key role in the purity of the metallic content, thus being one of the decisive

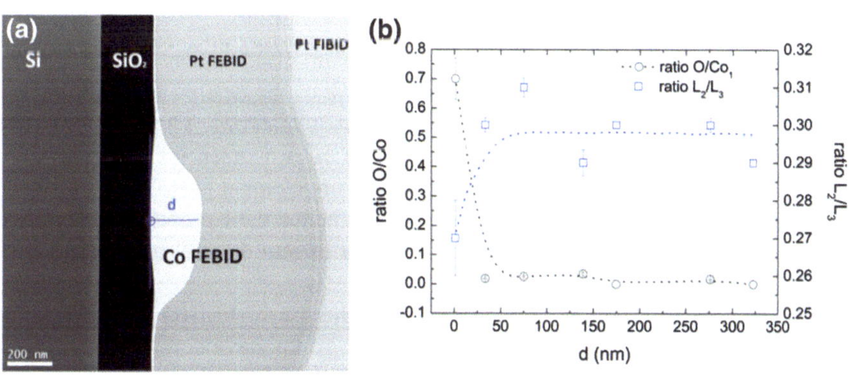

Fig. 3.6 **a** STEM-HAADF image of deposit 2. **b** Profiles of relative concentration of the O/Co and L_2/L_3 intensity ratios along the growth direction (*blue line* marked in **a**). Figure reprinted from [17]

parameters in the production of pure cobalt. The deposits grown at a high beam current (nanoamp range) have high cobalt content, whereas those grown at low beam currents (picoamp range) have low cobalt content, where a significant amount of oxidized cobalt together with metallic cobalt has been detected.

However, the FEBID process involves complex phenomena, and other relevant mechanisms have also been highlighted in the literature using different deposition parameters. For example, the autocatalysis effect [21] and the influence of the dwell time on the final composition [22] have been put forward. Thus, given a cobalt structure geometry, the cobalt content will be determined by the set of deposition parameters such as precursor flux, dwell time, refresh time, and beam current, and not only the beam current. Some of the deposition parameters relevant to the final content of the deposits will be studied in subsequent sections with significant results.

The strong differences in the microstructure and chemical nature of the deposits found in this systematic study might explain the different transport and magnetic properties reported in the literature for Co-based nanostructures grown by FEBID. Thus, under the same deposition conditions as chosen in the literature [9, 13, 20], samples grown at high I_e show metallic electrical transport and ferromagnetic behavior [9, 13], in sharp contrast with the semiconducting behavior exhibited by deposits grown at a low I_e [13].

We must emphasize that our work is the first study that clarifies the nature and electronic state of cobalt in the deposits grown by nanoscale FEBID, which has been addressed through the use of appropriate techniques: STEM, EELS and HRTEM.

3.4 Influence of the Substrate Temperature on the Nanostructure Composition: Co Pillars Grown at Low Current

As mentioned in Sect. 1.3.4 (Chap. 1), the main disadvantage of the FEBID technique is that the deposit does not have 100 % of the desired element, mainly due to the partial decomposition of the precursor material. The contaminating elements are usually carbon and oxygen, which often worsen the physical properties of the deposit compared with those of the pure element [23]. With the aim of improving the metal content of the deposits and hence their properties and future applications, various strategies, such as in situ or ex situ thermal treatments, have been reported [10].

However, one exception to this problem is the use of FEBID with $Co_2(CO)_8$ precursor gas. Several researchers, including our group, have devoted their efforts to the study of deposits using this precursor material, trying to explain the deposition mechanism, improve the metal content, and find applications in the field of nanotechnology; such as the fabrication of magnetic sensors, domain wall conduits, etc. [5–9, 13, 20, 22, 24–27]. In recent years, several researchers agreed on the need to generate heat in the deposition area to increase the metal content of the deposits, obtaining high purity deposits (>90 % Co); the latter was obtained by our group using high I_e, as remarked in Sect. 3.2. On the other hand, the use of high beam currents jeopardizes the lateral resolution of the Co nanodeposits because the diameter of the beam scales with the square root of the beam current [28]. Thus, it is highly desirable to use low FEB currents (picoamps) in order to grow Co nanostructures with high lateral resolution.

In the present work, we report the growth of Co nanostructures using low I_e (picoamps) with different substrate temperatures as well as the corresponding compositional analysis of the created Co nanopillars [29]. The aim of this study was to obtain high-purity Co nanostructures using low I_e (picoamps) as well as to clarify the influence of the local substrate heating effects on the Co content in the deposits. This experiment is feasible because the decomposition temperature of the $Co_2(CO)_8$ precursor material is relatively low [30, 31]. Ye et al. obtained the maximum growth rates by CVD at 140–200 °C at 1.3×10^{-5} mbar, temperatures easily achievable using a micro-hotplate.

The Co nanopillars were grown directly on the SiO_2 layer of the micro-hotplate set on "spot" mode with $V_e = 5$, 30 kV, $I_e = 25$, 44 pA, $P_{process} = 4 \times 10^{-6}$ mbar, $t_{process} = 5$ min, and variable $T_{substrate}$. We varied $T_{substrate}$ during each nanopillar growth using a commercial micro-hotplate (KMHP-100 from Kebaili Corporation) based on a Pt heater embedded in a sandwich-type suspended $SiO_2/Si_3N_4/SiO_2$ membrane, with a working area of about 250 μm × 150 μm (see Fig. 3.7), which we discussed extensively in Sect. 2.2 (Chap. 2).

We designed a sample holder with electrical connections to the micro-hotplate that are led from the process chamber to a DC voltage source, as discussed in Sect. 2.2 (Chap. 2). The micro-hotplate was calibrated under our vacuum conditions using the resistance values of the calibration carried out at atmospheric pressure by the manufacturer. As expected, it was necessary to apply a lower power to get the desired temperature.

Figure 3.8 presents SEM images of the nanopillars grown at 30 kV, 44 pA in (a) and at 5 kV, 25 pA in (b) with $T_{substrate}$ indicated. Observing the images we can see that the volume and diameter of the nanopillar's base increase with $T_{substrate}$ until it reaches 100 °C. Above this temperature, we observe an undesired effect, the growth of a large and anomalous halo around the nanopillar. The halo is a consequence of the FEBID technique, arising from the SEs produced by the BSEs when the FEB interacts with the substrate, where precursor gas molecules are adsorbed [32].

However, in our case the halo is larger in volume than usual, because together with SEs of 2nd order, there is the high $T_{substrate} > 100$ °C. Therefore, it seems that there is a decrease in the dissociation energy of the bonds in the molecule when $T_{substrate}$ is increased. A change in $T_{substrate}$ would lead to a change in the curve of cross section $\sigma_{(E)}$, whose maximum would shift to lower energies. Thus, it is expected that SEs that reach far away from the center of the pillar, and consequently have a low energy, are capable of dissociating gas molecules faster as $T_{substrate}$ increases.

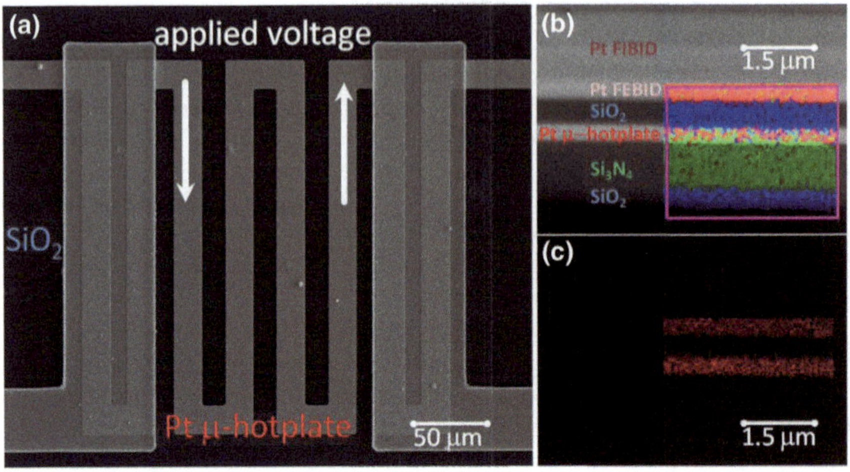

Fig. 3.7 **a** SEM image of the micro-hotplate (Kebaili) at low magnification with the stage at 0° tilt. Arrows indicate the Pt heater in which current flows, heating the suspended membrane. **b** SEM image at high magnification of a cross section of the micro-hotplate with the stage at 52° tilt. On top of it are grown two Pt layers, one by FEBID and the other one by FIBID, to protect the area and increase contrast. In the outlined area are indicated EDS spectra collected from this area, which clearly distinguish the micro-hotplate layers. **c** EDS mapping of the Pt spectrum, where the Pt layer of the micro-hotplate and the FEBID Pt layer were identified

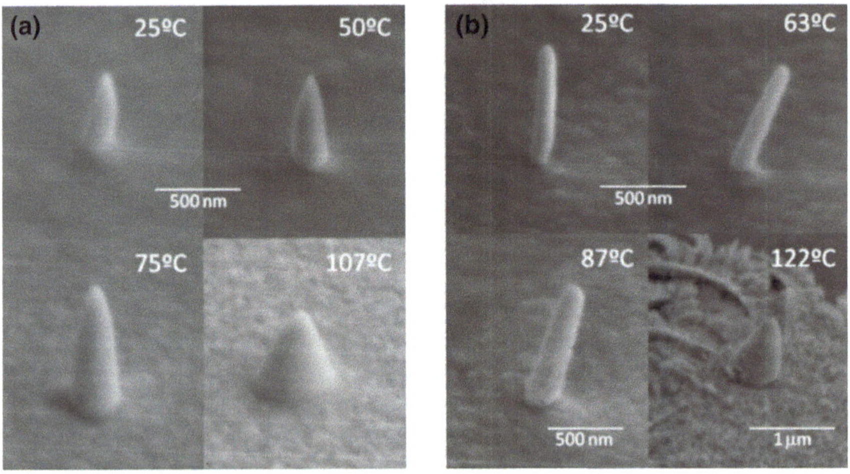

Fig. 3.8 SEM images (52° tilt) of Co nanopillars grown at **a** 30 kV and 44 pA and **b** at 5 kV and 25 pA with $T_{substrate}$ indicated. Figure reprinted with permission from [29]. Copyright from (2010), Elsevier

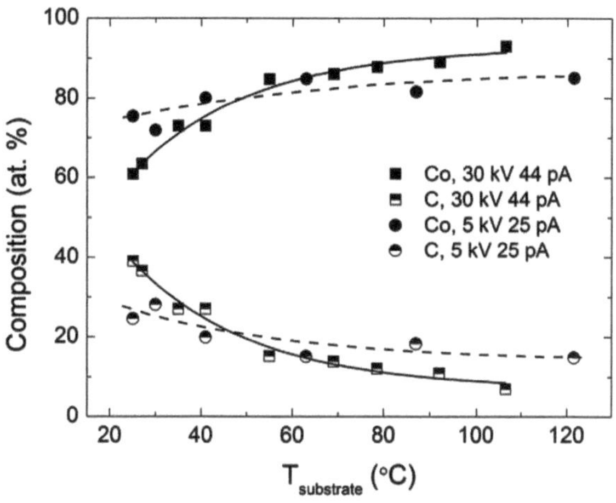

Fig. 3.9 Composition of Co nanopillars in atomic percent grown at 30 kV, 44 pA and 5 kV, 25 pA as a function of $T_{substrate}$. Figure adapted with permission from [29]. Copyright from (2010), Elsevier

Moreover, the height of the nanopillars increases as a function of $T_{substrate}$ up to ~100 °C, although it is known that the residence time of adsorbed molecules on the substrate surface decreases with $T_{substrate}$ and thus the growth rate should decrease with increasing $T_{substrate}$. This can be explained by the growth rate, which is expected to be limited by the gas flow. The nanopillar's height may increase with increasing $T_{substrate}$ owing to the modification of the cross-section curve. The total volume of the deposit should saturate at a given temperature when no more gas precursor is available. Detailed quantitative simulations would be needed to confirm this qualitative explanation, which would go beyond the scope of this study.

In Fig. 3.9 the Co content of the Co nanopillars as obtained with EDS measurements is shown as atomic percentage as a function of $T_{substrate}$. The EDS analysis was carried out at 9 kV, with a beam diameter of ~7.5 nm. The electron beam interaction at 9 kV in Co nanopillars was estimated using the CASINO program, which is based on Monte Carlo model simulations [33]. Thus, we see that the trajectories of the electrons accelerated at 9 kV that enter our nanopillars do not continue out of them, without needing a correction due to the contribution of the substrate. The Co content increases with temperature from 60 % at room temperature to 93 % at 107 °C in nanopillars fabricated at 30 kV, 44 pA and from 72 % at room temperature to 85 % at 122 °C in nanopillars grown at 5 kV, 25 pA. The Co content obtained at room temperature is in agreement with the results achieved in Sect. 3.2 and published in [13]. The results clearly show that when the substrate temperature rises, the nanopillars present a higher Co content. We do not plot the oxygen content because it is always below 2 %, its contribution being thus marginal in the reported results. Slight differences in the Co content between the nanopillars grown at 30 kV and 5 kV were found, probably owing to the different electron beam-induced heating in each case. We found $T_{substrate}$ ~ 100 °C is a good trade-off for high-purity cobalt nanostructures without noticeable changes in lateral dimensions.

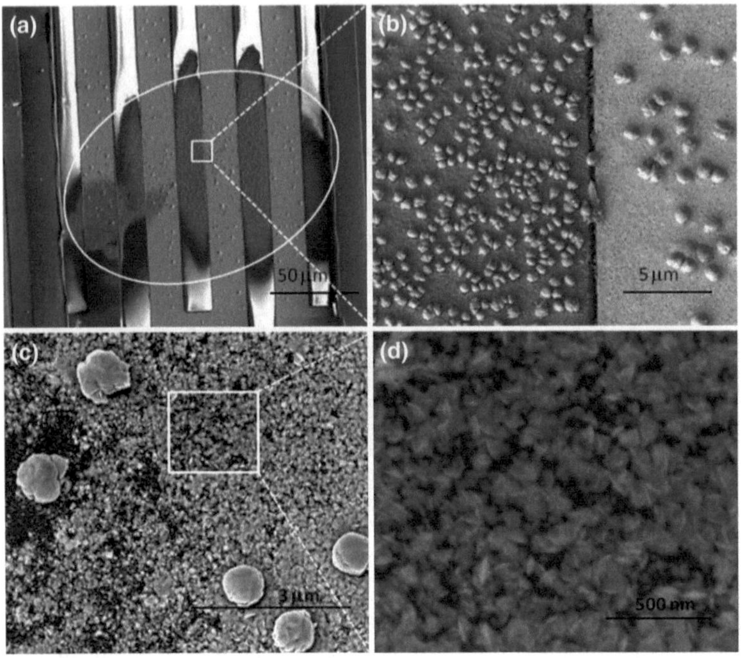

Fig. 3.10 SEM images of pure Co particles obtained by spontaneous thermal decomposition of $Co_2(CO)_8$ when it comes into contact with the micro-hotplate at ~197 °C. **a** The white ellipse indicates the area where the GIS delivers gas molecules onto the micro-hotplate and they immediately dissociate without FEB assistance. **b** The particle size depends slightly on the substrate, being greater if the Pt layer is underneath. **c**, **d** Cobalt particles and nanocrystals grown on the micro-hotplate with the Pt layer underneath. Figure reprinted with permission from [29]. Copyright from (2010), Elsevier

In our case, we note that the combination of local heat generated by the FEB and high $T_{\text{substrate}}$ are responsible for the decomposition of precursor gas and thus purify the nanodeposits.

This is further confirmed by the finding that a $T_{\text{substrate}}$ threshold exists at which thermal decomposition already occurs when the precursor gas molecules are in contact with the micro-hotplate without assistance from the electron beam. Under our vacuum and operation conditions, this temperature value is slightly below 200 °C. This value is consistent with maximum cobalt deposition rates between 140 and 200 °C obtained with CVD using $Co_2(CO)_8$ at 1.3×10^{-5} mbar [31]. Figure 3.10a shows the cobalt clusters and nanocrystals that appeared at ~197 °C where the impinging precursor molecules have been adsorbed on the micro-hotplate. In Fig. 3.10b it is possible to distinguish the different cluster sizes grown with and without the Pt microheater directly underneath, which presumably slightly modifies the surface temperature and the SE energy distribution on the surface. In the first case, the Co cluster size is around 800 nm. In the second case, it is around 600 nm. In Fig. 3.10c, d the Co clusters and nanocrystals (size below 100 nm) grown with the Pt microheater layers underneath can be observed. The in situ EDS measurements of these nanostructures indicated a cobalt content of ~98 at.%.

In order to clarify the importance of $T_{substrate}$ during deposition, we also performed in situ post-deposition thermal treatments as previously reported for Pt NWs [34]. Co NWs with dimensions 1.2 μm × 250 nm × 100 nm (length × width × thickness) were grown at 30 keV and 44 pA at room temperature. The Co content was 60 at.% (atomic) as mentioned before for these growth conditions (Fig. 3.9). The temperature of the micro-hotplate was then increased slowly up to 600 °C and maintained at this value for 40 min. The in situ EDS measurement of the Co nanowire after this thermal annealing did not reveal any change in the composition. This experiment evidences the relevance of the combination of the FEB and the high substrate temperature *during the deposition process* to obtain pure cobalt nanostructures. Once the deposit is fabricated on the substrate, the carbon matrix cannot be decomposed by heat treatment at 600 °C under our vacuum conditions.

3.5 Autocatalytic Growth of Cobalt Using Dicobalt Octacarbonyl

In the literature a specific effect, called the autocatalytic effect, has been reported, in which the precursor gas molecules decompose, dissociating the bonds between metal and ligands autocatalytically and thus allowing the growth of pure metals without FEB irradiation. This effect is usually produced at room temperature. Some examples of it have been observed in ultrahigh vacuum (UHV) systems, $P_{base} \sim 10^{-10}$ mbar using $Fe(CO)_5$ [35–37], Si_2H_6, and WF_6 [37–39] as precursor meterials. Some researchers went further and spoke about FEB scanning as an invisible ink that exposes a pattern on SiO_2 without a gaseous precursor. Apparently, the FEB exposes the scanned area and then releases the gas precursor, which selectively decomposes autocatalytically, creating pure Fe structures in the place where FEB has scanned. However, the cause of the initial dissociation of the one molecule that starts the dissociation of the rest of them is not clear. The hypothesis is that the activation process may be a consequence of the formation of oxygen vacancies generated by the FEB-induced desorption, and the SiO_2 is reduced to SiO, and this could be the active species that causes initial decomposition of the precursor.

These results were the inspiration for our study using $Co_2(CO)_8$ as a precursor material. However, the first setback we found is that our system works under high vacuum (HV); the pressure in the process chamber is typically about 1×10^{-6} mbar. Thus, we must keep in mind that the process chamber has contaminating species such as water vapor and hydrocarbon molecules, which are adsorbed on the substrate surface and are deposited by the FEB, creating a protective barrier. Therefore, it is very difficult for molecules of a precursor material such as $Co_2(CO)_8$ to decompose autocatalytically on the surface where the electron beam has scanned in advance in a system that operates at HV.

A series of experiments on various types of substrates were carried out to study whether $Co_2(CO)_8$ exhibits a catalytic or autocatalytic effect. Substrate A: Si; substrate B: Si + Co FEBID; substrate C: Si + Co layer (20 nm) + Al layer (2 nm), both grown by sputtering.

Experiments on substrate A were: (1) To expose the Si substrate (previously cleaned with acetone, alcohol, and water in an ultrasonic bath, and without FEB irradiation) to an atmosphere of precursor gas ($P_{process} = 1 \times 10^{-5}$ mbar) for 50 min. (2) FEB irradiation at 5 kV and 86 pA, lines <100 nm in width, $t = 5$ min, total dose = 258 nC/μm^2 on Si substrate. Subsequently, the substrate was exposed to an atmosphere of precursor gas ($P_{process} = 1 \times 10^{-5}$ mbar) for 50 min. Further inspection of the surface where the autocatalytic reaction should occur because of the orientation of the GIS shows no Co growth in either case. Therefore, we conclude that there is no autocatalytic effect on substrate A.

The experiment on substrate B was the growth of Co FEBID pillars to be used as seed for the subsequent activation or autocatalytic decomposition of the precursor. Co pillars were grown using currents in the picoamp range as well as in the nanoamp range in order to study whether the Co content has an effect on catalytic decomposition. The Co content in these pillars ranges from 60 to 90 %. After the growth of the seeded nanostructures, they were exposed to a precursor gas atmosphere ($P_{process} = 1 \times 10^{-5}$ mbar) for 50 min. Inspection of the Co pillars grown by FEBID does not show any growth of Co thereafter. Thus, we conclude that the catalytic effect is not produced on substrate B either.

On substrate C, we controlled the milling depth by FIB of Co and Al layers grown by sputtering. We performed an etching of 10 squares by FIB of dimensions 1 μm × 1 μm × (3–25) nm ($x \times y \times z$) using 30 kV and 10 pA. First, we milled 2 nm of Al layer and then 20 nm of Co layer until the Si substrate was reached. After that, these structures were exposed to a gas precursor atmosphere for 30 min. The pressure in the chamber varies from $P_{base} = 1 \times 10^{-6}$ mbar to $P_{process} = 1 \times 10^{-5}$ mbar. Thus, we studied when the catalytic decomposition effect starts on the Co layer and used it as a seed.

Figure 3.11 shows SEM images as examples of the study on substrate C. (a) shows a SEM image of the surface after being irradiated by a FIB of volume 1 μm × 1 μm × 0.003 μm and (b) shows a SEM image of the surface after being irradiated by a FIB of volume 1 μm × 1 μm × 0.003 μm and 30 min of the precursor gas atmosphere. We can see the growth of Co nanocrystals ~5 nm in size, which form agglomerates of about 40 nm. The composition analysis by EDS confirmed that Co is pure. The estimated growth rate of the Co nanocrystals is 1.33 nm/min. (c) Shows a SEM image of the surface (1 μm × 1 μm × 0.007 μm) after being irradiated by a FIB and (d) shows a SEM image of the surface after being irradiated by a FIB of volume 1 μm × 1 μm × 0.007 μm and 30 min of the

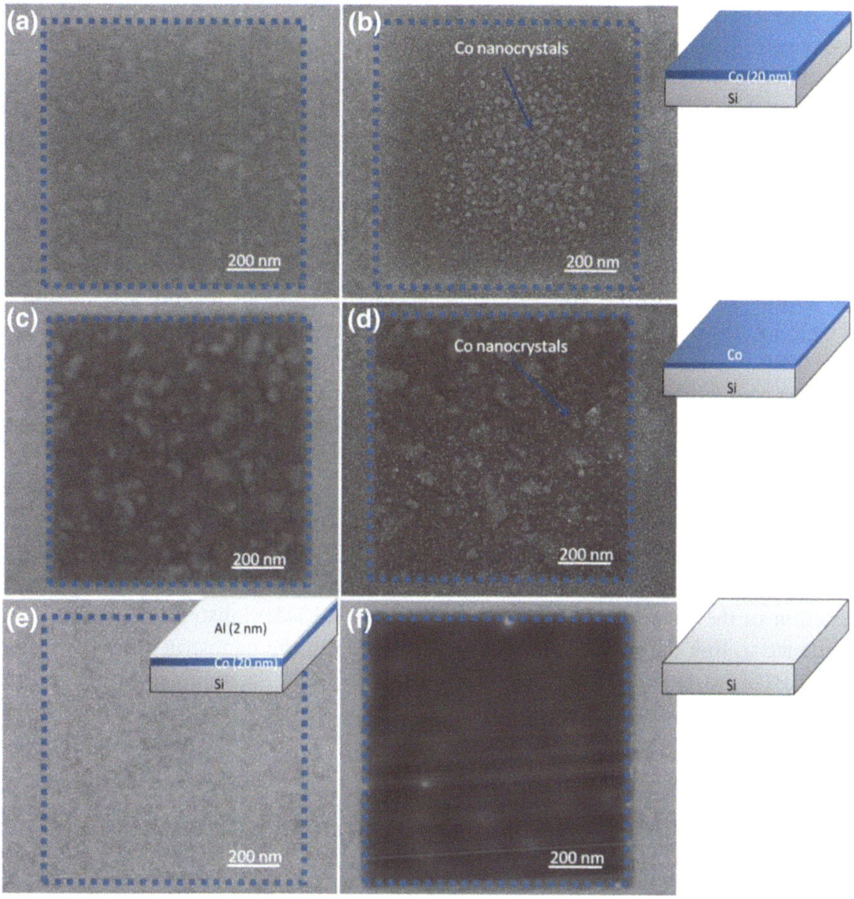

Fig. 3.11 SEM images of the Co nanocrystals obtained by autocatalytic growth on top of a Co surface revealed after FIB milling of substrate C. **a** SEM image of substrate C after FIB milling (1 μm × 1 μm × 0.003 μm); **b** SEM image of substrate C after FIB milling (1 μm × 1 μm × 0.003 μm) and under precursor gas atmosphere for 30 min; **c** SEM image of substrate C after FIB milling (1 μm × 1 μm × 0.007 μm); **d** SEM image of substrate C after FIB milling (1 μm × 1 μm × 0.007 μm) and under precursor gas atmosphere for 30 min; **e** SEM image of substrate C; **f** SEM image of substrate C after FIB milling (1 μm × 1 μm × 0.025 μm) and under precursor gas atmosphere for 30 min. Figure adapted with permission from [40]. Copyright from (2012), Elsevier

precursor gas atmosphere. This image shows the appearance of small Co nanocrystals ~5 nm in the irradiated area. However, note that both the size and the number of nanocrystals is less than in (b). The estimated growth rate is 0.17 nm/min, significantly lower than the growth rate in (b). This effect can be explained by the high Ga dose on the substrate implanting therein and modifying the Co surface. Thus, the contamination of the surface by Ga ions efficiently prevents the decomposition of the precursor gas. In (e), we show a SEM image of the surface of the virgin substrate. In (f), we show a SEM image of the surface after being irradiated by FIB, 1 μm × 1 μm × 0.025 μm, where the Co layer completely was eliminated and the substrate exposed to 30 min of the precursor gas atmosphere was silicon. We did not observe the appearance of small Co nanocrystals, so we have proof that the catalytic effect occurs only if there is a seed.

The conclusions of this study are that under vacuum ($P_{base} \sim 1 \times 10^{-6}$ mbar) and our working conditions, there is no autocatalytic or catalytic effect on the substrates A (Si), B (Si/Co FEBID), or C (Si/Co/Al), if the Al layer is not milled. In A, this is due to the contaminating species present in the process chamber, which are adsorbed and deposited on the substrate surface when this is irradiated by a FEB, creating a protective barrier. In B, we used Co FEBID nanopillars as a catalyst to the reaction, which have a high Co percentage but are not pure Co. So, it seems essential for the reaction that the initial catalyst is pure Co. Experiments and results on substrate C are detailed in Fig. 3.11. On substrate C, a catalytic effect is observed, as we can see by the growth of Co nanocrystallites of 5 nm size (see Fig. 3.11b–d). The decomposition of the precursor on the Co grown by sputtering occurs until we mill through to the Si substrate (Fig. 3.11f), when a catalytic effect is no longer observed because the initial seed for the decomposition reaction of the precursor is not present. These results demonstrate that there is no evidence to link the high content of Co ($\sim$90 %) in the nanostructures fabricated by FEBID presented in this chapter with the catalytic decomposition of the precursor gas ($Co_2(CO)_8$).

3.6 Ultranarrow Cobalt Nanostructures and Applications

In this section, we describe the experiments carried out to obtain pure Co nanostructures using low electron beam current at room temperature. As described in Sects. 3.2–3.4 as other authors have also confirmed [22], the tuning of some growth parameters allows us to modify the Co content of the deposits created by this technique. Following this approach, we fix a low beam current (in the picoamp range), as desired to obtain narrow structures, and study the influence of the precursor gas flux on the Co content.

3.6.1 Influence of the Precursor Gas Flow

In the study of the influence of the precursor gas flux on the Co content, cobalt cubes with sides of 100 nm and 100 nm thick were grown on silicon. The precursor flux can be tailored by gradually opening the gas valve, and the chamber pressure will change accordingly and will be monitored. In situ compositional analysis was carried out on the cubes by EDS at 3 kV.

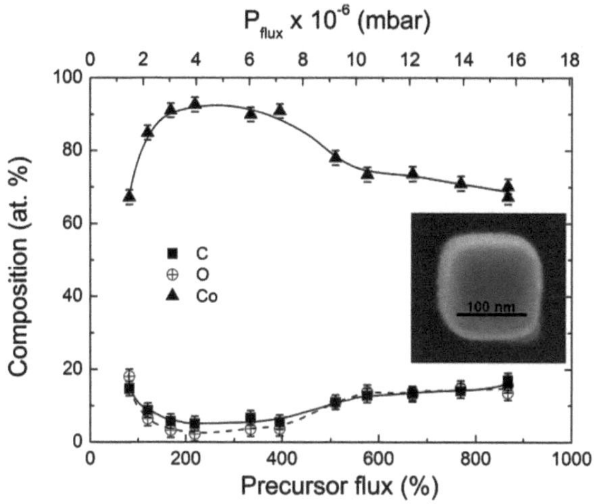

Fig. 3.12 Co atomic percentage in the nanodeposits grown at 3 kV and 21 pA as a function of the $Co_2(CO)_8$ precursor flux. A SEM image of the typical cobalt cubes grown is shown in the inset. The relative change of precursor flux is here defined by the expression: *Precursor flux* [%] = 100 × P_{flux}/P_{base}, where $P_{flux} = P_{process} - P_{base}$. Figure adapted with permission from [41]. Copyright from (2011), ACS

The deposition parameters for the growth of the Co cubes are $V_e = 3$ kV; $I_e = 21$ pA; $P_{base} = 1.3 \times 10^{-6}$ mbar; $t_{process} = 103$ s; $P_{process}$ = variable. The cobalt content presents a clear trend as a function of the precursor gas flux as can be seen in Fig. 3.12. The cobalt content that is usually found at low beam current, around 65 %, is obtained with the lowest and the highest precursor flux used. However, at intermediate precursor flux values, the cobalt content of the nano-cubes can be optimized up to ~93 %. These results can be understood by taking into account the different deposition regimes explained in detail in Sect. 1.3.4. At low precursor flux, where the growth is performed within the precursor-limited regime, dissociation of residual species present in the process chamber will play a role, producing sample contamination [22]. At high precursor flux, where the growth is performed within the electron-limited regime, the precursor molecules are refreshed too quickly for the electron beam current to decompose completely all the precursor material, which will become incorporated in the nanodeposit.

At intermediate precursor flux, where the growth is performed between the two regimes, it is possible to reach a perfect balance between the precursor flux and the electron beam current. It is thus possible to produce the total decomposition of the precursor material while the incorporation of residual species is avoided. In those deposition conditions, it is thus feasible to reach narrow Co nanostructures <50 nm in size and with high purity, >90 at.%.

Results obtained in the study of the precursor gas flow are suitable for Co nanostructures with dimensions close to 100 nm and using the same deposition parameters. In other conditions, additional experiments should be performed to find the suitable precursor gas flow. Other relevant parameters to tune the Co content in the nanostructures are DT, RT, and dimensions of the structures, as already noted by other researchers [21, 22, 42, 43] and our research group [41].

3.6.2 Ultranarrow Cobalt Nanostructures

Once we had mastered the growth conditions for low beam current and high Co, we focused our attention on the growth of two kinds of high-purity narrow cobalt nanostructures (see Fig. 3.13). Type 1: High-purity ultranarrow cobalt nanowires (linear and L shaped), with prospective applications related to spintronics [24] and magnetic logic based on domain wall manipulation [20]. The corresponding SEM images are displayed in Fig. 3.13a, b, showing an impressive high lateral resolution, around 30 nm. Type 2: Cobalt nanometer Hall crosses for the fabrication of nanosensors [24, 25], as shown in Fig. 3.13c, d.

Type 1 ultranarrow nanostructures were grown at 3 kV and 21 pA. The composition of these is 90:8:2 (Co:C:O) in atomic percent. The lateral resolution achieved is 30 nm, nearly four times better than in previous studies in which it was necessary to use I_e in the nanoamp range to obtain high-Co-content deposits (>90 %) [13]. Magnetic characterization of this small structures can be carried out

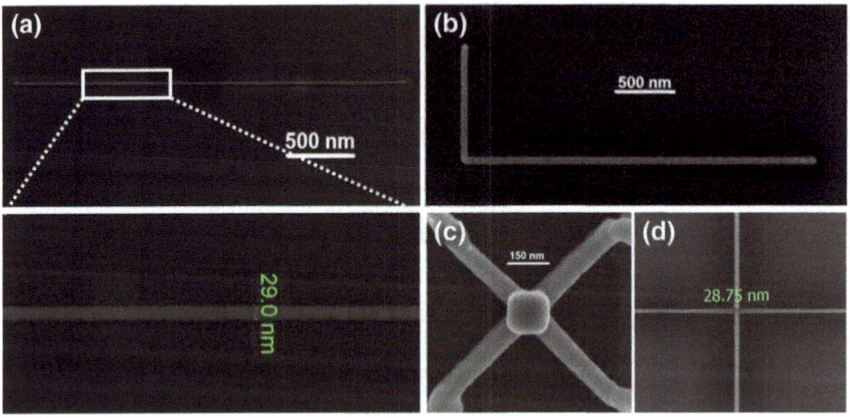

Fig. 3.13 SEM images of cobalt nanostructures grown at 3 kV, 21 pA. **a** NW with dimensions 29 nm (*width*), 3 μm (*length*), and 30 nm (*thickness*) (image taken at tilting angle of 52°). A higher-magnification image of the outlined area is shown in the panel below **a** (image taken at 0° tilting angle). **b** Nanostructure (L shaped) with a 3 μm long horizontal part and 1 μm long perpendicular part, all 40 nm wide and 30 nm thick (image taken at 0° tilting angle). **c** Cobalt Hall effect sensor based on four independent 75 nm-wide cobalt lines and the central square 150 nm-wide active area (image taken at 0° tilting angle). **d** Cobalt nano–Hall effect sensor based on two 30 nm-wide cobalt lines (image taken at 0° tilting angle). Figure reprinted with permission from [41]. Copyright from (2011), ACS

by TEM techniques such as electronic holography [44–47]. A study of electronic holography in ultranarrow NWs was performed on a Titan [3] High-base with a corrector for spherical aberration in the objective lens. The magnetic induction obtained by electronic holography in NWs 30 nm wide and 9 nm thick was 1.8 T, which agrees well with the saturation magnetization of pure Co. NWs with different thickness have been studied; in some cases the magnetic induction decreases by about 20 %, probably due to small changes in the NWs' composition. Thus, quantitative magnetic measurements by electronic holography demonstrate magnetism and functionality in Co ultranarrow NWs for applications in spintronics [41].

With regard to type 2 ultranarrow nanostructures, small Hall effect sensors exhibit increased detection sensitivity, expressed as magnetic flux values below $10^{-6}\phi_0/\text{Hz}^{-1/2}$ ($\phi_0 = h/2e$, where h is Planck's constant and e is the electronic charge, f_0 is a flux quantum) [25]. The fabrication of such devices on the nanometer scale has aroused strong interest in the field of biosensors for detecting biological fractions conjugated to magnetic nanoparticles and in general applications in nanosensing, such as scanning Hall microscopy, nanomagnetometry, current sensors, and proximity switches. [48, 49].

3.6.3 Cobalt Nano–Hall Effect Sensors

Cobalt nano–hall effect sensors were grown at 3 kV and 21 pA in the appropriate precursor flux and with suitable dwell time on a thermally oxidized silicon substrate, which was first micropatterned by optical lithography techniques in order to produce Ti contact pads.

First, Pt FIBID NWs were grown to allow the subsequent connection of the Hall effect sensors to the titanium pads as observed in Fig. 3.14. In the next step, four independent cobalt nanowires were grown in cross shape with 40–75 nm width. Finally, a cobalt square with a typical width of 150 nm or less (the active area of the Hall effect sensor) was grown in the middle of the four nanowires, closing the electrical circuit. EDS analysis of the sensing areas indicated percentages of cobalt close to ~90 %.

The Hall effect (HE) measurements using the 4-probe geometry were performed in a commercial physical property measurement system (PPMS) in the temperature range from 300 K down to 2 K. The applied magnetic field is always perpendicular to the substrate from 9 to –9 T. The bias current used in the measurements ranged from 10 μA to 1.2 mA. In Fig. 3.15, the Hall resistivity of the sensor displayed in Fig. 3.13c is represented as a function of the applied magnetic field at 300 and 10 K. The Hall resistivity is given by Eq. (1.4).

The lines in Fig. 3.15 correspond to the individual contributions from the ordinary Hall effect (OHE) and the anomalous Hall effect (AHE). The crossover from the low-field behavior, positive in sign and dominated by the AHE, to the high-field behavior, negative in sign and dominated by the OHE, takes place at ~2 T, as occurs in bulk Co [44] and in Co FEBID grown at high beam current [13].

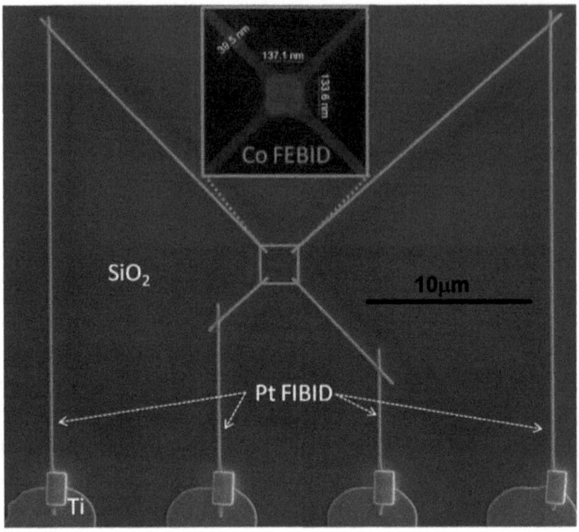

Fig. 3.14 SEM images of the fabrication process of cobalt nano–Hall effect sensors. First, four Pt FIBID NWs were grown, then, in the blue square indicated, four Co FEBID NWs were grown, and finally the Co FEBID square was grown

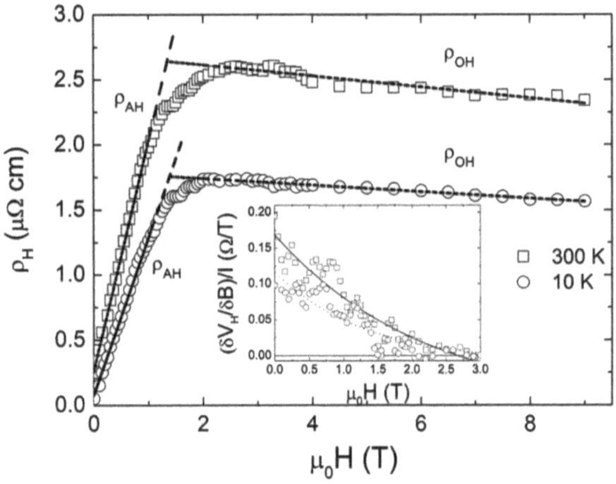

Fig. 3.15 Hall resistivity in the same Co Hall effect sensor as shown in Fig. 3.13c at 300 and 10 K. The inset shows the field sensitivity as a function of the magnetic field at 300 and 10 K. Figure adapted with permission from [41]. Copyright from (2011), ACS

It should be pointed out that low-Co-content Hall effect sensors grown by FEBID do not present this clear crossover between the AHE and the OHE [24, 25].

The field sensitivity (S_I) is an important parameter to characterize Hall effect sensors. From

$$S_I = \frac{\delta V_H/\delta B}{I} \tag{3.1}$$

S_I is calculated and represented as a function of the applied magnetic field in the inset of Fig. 3.15.

The low-field S_I of the studied Hall effect sensors ranges from 0.13 to 0.07 Ω/T for contents of 88:8:4 and 84:9:7 (Co:C:O) (atomic percent), respectively. The

Table 3.3 Composition of cobalt nanodeposits grown on different types of substrate at 3 kV, 21 pA

Substrate	C (at.% ± 2 %)	O (at.% ± 2 %)	Co (at.% ± 2 %)
Si	6	10	85
Si_3N_4	7	8	85
SiO_2	2	7	91

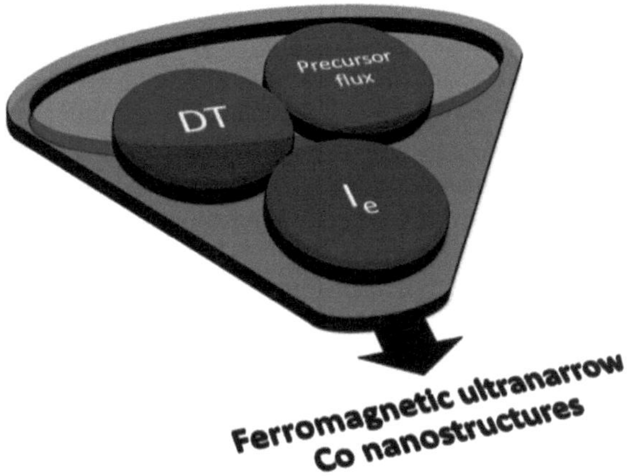

Fig. 3.16 Sketch of the main deposition parameters that affect the formation of ferromagnetic ultranarrow Co nanostructures

obtained S_I values are similar to those reported in the literature [24] for lower-content Co sensors but the important figure of merit of nanoscale Hall effect sensing is the minimum magnetic flux detectable (ϕ_{min}). In Ref. [25], ϕ_{min} is estimated to be around 4.5×10^{-6} ϕ_0, where ϕ_{min} is the flux quantum, which is better than previously reported values in other nano–Hall probes. In the thermal-noise regime, ϕ_{min} is proportional to the square root of the longitudinal sensor resistance and the sensing area. As both the longitudinal resistance and the sensing area are significantly lower in our high-purity narrow Co sensors than in Ref. [25], we expect to the ϕ_{min} to be improved by more than one order of magnitude. Dedicated experiments to confirm this issue are in progress, which constitutes an exciting future research line in nanoscale sensing.

In the experiments reported in this section, Co nanostructures were grown on three different substrates: silicon, silicon oxide (~250 nm) on silicon, and silicon nitride membranes 50 nm thick. As shown in Table 3.3, the Co concentration does not vary significantly in nanodeposits grown using the same deposition parameters.

We have also taken a big step toward understanding the mechanism of growth by FEBID, and more specifically the mechanism involved for this precursor material, $Co_2(CO)_8$. Our aim was to find suitable growth parameters to obtain ferromagnetic ultranarrow Co nanostructures (see Fig. 3.16).

3.7 Conclusions

In this section, we have studied in detail Co deposits grown by FEBID using $Co_2(CO)_8$ as a precursor material by means of different approaches such as EDS, STEM, HRTEM, EELS, electronic holography, and magnetotransport measurements.

Additionally, we have varied in a controlled way some deposition parameters that can significantly improve the Co content, such as I_e, the substrate temperature, the nature of the substrate, and the precursor flow. The Co content has a major influence on their magnetic properties and therefore on their future applications in logic, sensing, and magnetic storage. The growth of ferromagnetic nanostructures with 30 nm width is a great advance for the functionality of the FEBID technique. However, we are only at the early stages of this project, in which the growth of structures below 10 nm size [50] and the fabrication of three-dimensional ferromagnetic structures [51] can be achieved.

References

1. Lin, T., Tsang, C., Fontana, R.E., Howard, J.K.: Exchange-coupled Ni-Fe/Fe-Mn, Ni-Fe/Ni-Mn and NiO/Ni-Fe films for stabilization of magnetoresistive sensors. IEEE Trans. Magn. **31**(6), 2585–2590 (1995)
2. Chappert, C., Fert, A., Van Dau, F.N.: The emergence of spin electronics in data storage. Nat. Mater. **6**(11), 813–823 (2007)
3. Allwood, D.A., Xiong, G., Faulkner, C.C., Atkinson, D., Petit, D., Cowburn, R.P.: Magnetic domain-wall logic. Science **309**(5741), 1688–1692 (2005)
4. Martins, V.C., Germano, J., Cardoso, F.A., Loureiro, J., Cardoso, S., Sousa, L., Piedade, M., Fonseca, L.P., Freitas, P.P.: Challenges and trends in the development of a magnetoresistive biochip portable platform. J. Magn. Magn. Mater. **322**(9–12), 1655–1663 (2008)
5. Utke, I., Hoffmann, P., Berger, R., Scandella, L.: High-resolution magnetic Co supertips grown by a focused electron beam. Appl. Phys. Lett. **80**(25), 4792–4794 (2002)
6. Utke, I., Bret, T., Laub, D., Buffat, P., Scandella, L., Hoffmann, P.: Thermal effects during focused electron beam induced deposition of nanocomposite magnetic-cobalt-containing tips. Microelectron. Eng. **73–74**, 553–558 (2004)
7. Utke, I., Michler, J., Gasser, P., Santschi, C., Laub, D., Cantoni, M., Buffat, P.A., Jiao, C., Hoffmann, P.: Cross section investigations of compositions and sub-structures of tips obtained by focused electron beam induced deposition. Adv. Eng. Mater. **7**(5), 323–331 (2005)
8. Lau, Y.M.; Chee, P.C.; Thong, J.T.L.; Ng, V.: Properties and applications of cobalt-based material produced by electron-beam-induced deposition. J. Vac. Sci. Technol. A—Vac. Surf. Films **20**(4), 1295–1302 (2002)
9. Fernández-Pacheco, A., De Teresa, J.M., Szkudlarek, A., Córdoba, R., Ibarra, M.R., Petit, D., O'Brien, L., Lewis, E.R., Read, D.E., Zeng, H.T., Cowburn, R.P.: Magnetization reversal in individual cobalt micro- and nanowires grown by focused-electron-beam-induced-deposition. Nanotechnology **20**, 475704 (2009)
10. Botman, A.; Mulders, J.J.L.; Hagen, C.W., Creating pure nanostructures from electron-beam-induced deposition using purification techniques: a technology perspective. Nanotechnology **20**(37), 372001 (2009)
11. Gazzadi, G.C., Mulders, J.J.L., Trompenaars, P., Ghirri, A., Rota, A., Affronte, M., Frabboni, S.: Characterization of a new cobalt precursor for focused beam deposition of magnetic nanostructures. Microelectron. Eng. **88**(8), 1955–1958 (2011)

12. Gazzadi, G.C., Mulders, H., Trompenaars, P., Ghirri, A., Affronte, M., Grillo, V., Frabboni, S.: Focused electron beam deposition of nanowires from cobalt tricarbonyl nitrosyl (Co(CO)(3)NO) precursor. J. Phys. Chem. C **115**(40), 19606–19611 (2011)
13. Fernández-Pacheco, A.; De Teresa, J.M.; Córdoba, R.; Ibarra, M.R.: Magnetotransport properties of high-quality cobalt nanowires grown by focused-electron-beam-induced deposition. J. Phys. D: Appl. Phys. **42**(5), 055005 (2009)
14. Jeanguillaume, C., Colliex, C.: Spectrum-image—the next step in EELS digital acquisition and processing. Ultramicroscopy **28**(1–4), 252–257 (1989)
15. Gloter, A.: Private communication
16. Trebbia, P., Bonnet, N.: EELS elemental mapping with unconventional methods. 1. Theoretical basis—image-analysis with multivariate-statistics and entropy concepts. Ultramicroscopy **34**(3), 165–178 (1990)
17. Córdoba, R., Fernández-Pacheco, R., Fernández-Pacheco, A., Gloter, A., Magén, C., Stephan, O., Ibarra, M.R., De Teresa, J.M.: Nanoscale chemical and structural study of Co-based FEBID structures by STEM-EELS and HRTEM. Nanoscale Res. Lett. **6**, 1–6 (2011). doi:10.1186/1556-276X-6-592
18. Mitterbauer, C., Kothleitner, G., Grogger, W., Zandbergen, H., Freitag, B., Tiemeijer, P., Hofer, F.: Electron energy-loss near-edge structures of 3d transition metal oxides recorded at high-energy resolution. Ultramicroscopy **96**(3–4), 469–480 (2003)
19. Golla-Schindler, U., Benner, G., Putnis, A.: Laterally resolved EELS for ELNES mapping of the Fe L-2, L-3- and OK-edge. Ultramicroscopy **96**(3–4), 573–582 (2003)
20. Fernández-Pacheco, A.; De Teresa, J.M.; Córdoba, R.; Ibarra, M.R.; Petit, D.; Read, D.E.; O'Brien, L.; Lewis, E.R.; Zeng, H.T.; Cowburn, R.P.: Domain wall conduit behavior in cobalt nanowires grown by focused electron beam induced deposition. Appl. Phys. Lett. **94**(19), 192509 (2009)
21. Utke, I., Golzhauser, A.: Small, minimally invasive, direct: Electrons induce local reactions of adsorbed functional molecules on the nanoscale. Angew. Chem. Int. Ed. **49**(49), 9328–9330 (2010)
22. Bernau, L., Gabureac, M., Erni, R., Utke, I.: Tunable nanosynthesis of composite materials by electron-impact reaction. Angew. Chem. Int. Ed. **49**(47), 8880–8884 (2010)
23. De Teresa, J.M., Córdoba, R., Fernández-Pacheco, A., Montero, O., Strichovanec, P., Ibarra, M.R.: Origin of the difference in the resistivity of as-grown focused-ion- and focused-electron-beam-induced Pt nanodeposits. J. Nanomaterials **2009**, 936863 (2009)
24. Boero, G.; Utke, I.; Bret, T.; Quack, N.; Todorova, M.; Mouaziz, S.; Kejik, P.; Brugger, J.; Popovic, R.S.; Hoffmann, P.: Submicrometer hall devices fabricated by focused electron-beam-induced deposition. Appl. Phys. Lett. **86**(4), 042503 (2005)
25. Gabureac, M.; Bernau, L.; Utke, I.; Boero, G.: Granular Co-C nano-hall sensors by focused-beam induced deposition. Nanotechnology **21**(11), 115503 (2010)
26. Belova, L.M.; Dahlberg, E.D.; Riazanova, A.; Mulders, J.J. L.; Christophersen, C.; Eckert, J.: Rapid electron beam assisted patterning of pure cobalt at elevated temperatures via seeded growth. Nanotechnology **22**(14), 145305 (2011)
27. Mulders, J.J.L.; Belova, L.M.; Riazanova, A.: Electron beam induced deposition at elevated temperatures: compositional changes and purity improvement. Nanotechnology **22**(5), 055302 (2011)
28. Molhave, K., Madsen, D.N., Rasmussen, A.M., Carlsson, A., Appel, C.C., Brorson, M., Jacobsen, C.J.H., Boggild, P.: Solid gold nanostructures fabricated by electron beam deposition. Nano Lett. **3**(11), 1499–1503 (2003)
29. Córdoba, R., Sesé, J., De Teresa, J.M., Ibarra, M.R.: High-purity cobalt nanostructures grown by focused-electron-beam-induced deposition at low current. Microelectron. Eng. **87**(5–8), 1550–1553 (2010). doi:10.1016/j.mee.2009.11.027
30. Kunz, R.R., Mayer, T.M.: Catalytic growth-rate enhancement of electron-beam deposited iron films. Appl. Phys. Lett. **50**(15), 962–964 (1987)

31. Ye, D.X., Pimanpang, S., Jezewski, C., Tang, F., Senkevich, J.J., Wang, G.C., Lu, T.M.: Low temperature chemical vapor deposition of Co thin films from $Co_2(CO)_8$. Thin Solid Films **485**(1–2), 95–100 (2005)
32. van Dorp, W.F.; Hagen, C.W.: A critical literature review of focused electron beam induced deposition. J. Appl. Phys. **104**(8), 081301 (2008)
33. Drouin, D., Couture, A.R., Joly, D., Tastet, X., Aimez, V., Gauvin, R.: CASINO V2.42—A fast and easy-to-use modeling tool for scanning electron microscopy and microanalysis users. Scanning **29**(3), 92–101 (2007)
34. Liao, Z.M.; Xu, J.; Zhang, X.Z.; Yu, D.P.: The relationship between quantum transport and microstructure evolution in carbon-sheathed Pt granular metal nanowires. Nanotechnology **19**(30), 305402 (2008)
35. Lukasczyk, T., Schirmer, M., Steinruck, H.P., Marbach, H.: Electron-beam-induced deposition in ultrahigh vacuum: Lithographic fabrication of clean iron nanostructures. Small **4**(6), 841–846 (2008)
36. Lukasczyk, T., Schirmer, M., Steinrueck, H.-P., Marbach, H.: Generation of clean iron structures by electron-beam-induced deposition and selective catalytic decomposition of iron pentacarbonyl on Rh(110). Langmuir **25**(19), 11930–11939 (2009)
37. Walz, M.-M., Schirmer, M., Vollnhals, F., Lukasczyk, T., Steinrueck, H.-P., Marbach, H.: Electrons as "Invisible Ink": fabrication of nanostructures by local electron beam induced activation of SiO_x. Angew. Chem. Int. Ed. **49**(27), 4669–4673 (2010)
38. Hirose, F.; Sakamoto, H.: Low-temperature Si selective epitaxial growth using electron-beam-induced reaction. Jpn. J. Appl. Phys. Part 1—Regul. Pap., Short Notes Rev. Pap. **34**(11), 5904–5907 (1995)
39. Matsui, S., Ichihashi, T., Mito, M.: Electron-beam induced selective etching and deposition technology. J. Vac. Sci. Technol., B **7**(5), 1182–1190 (1989)
40. Córdoba, R., Sesé, J., Ibarra, M.R., De Teresa, J.M.: Autocatalytic growth of Co on pure Co surfaces using $Co_2(CO)_8$ precursor. Appl. Surf. Sci. **263**, 242–246 (2012). doi:10.1016/j.apsusc.2012.09.037
41. Serrano-Ramón, L., Córdoba, R., Rodriguez, L.A., Magén, C., Snoeck, E., Gatel, C., Serrano, I., Ibarra, M.R., De Teresa, J.M.: Ultrasmall functional ferromagnetic nanostructures grown by focused electron-beam-induced deposition. ACS Nano **5**(10), 7781–7787 (2011). doi:10.1021/nn201517r
42. Beaulieu, D., Ding, Y., Wang, Z.L., Lackey, W.J.: Influence of process variables on electron beam chemical vapor deposition of platinum. J. Vac. Sci. Technol., B **23**(5), 2151–2159 (2005)
43. Kohlmann, K.T., Buchmann, L.M., Brunger, W.H.: Repair of open stencil masks for ion projection lithography by e-beam induced metal-deposition. Microelectron. Eng. **17**(1–4), 427–430 (1992)
44. Kotzler, J.; Gil, W.: Anomalous hall resistivity of cobalt films: Evidence for the intrinsic spin-orbit effect. Phys. Rev. B **72**(6), 060412 (2005)
45. Nagaosa, N., Sinova, J., Onoda, S., MacDonald, A.H., Ong, N.P.: Anomalous Hall effect. Rev. Mod. Phys. **82**(2), 1539–1592 (2010)
46. Kirk, K.J., Chapman, J.N., Wilkinson, C.D.W.: Lorentz microscopy of small magnetic structures (invited). J. Appl. Phys. **85**(8), 5237–5242 (1999)
47. Tonomura, A.: Electron holography of magnetic-materials and observation of flux-line dynamics. Ultramicroscopy **47**(4), 419–424 (1992)
48. Moritz, J.; Rodmacq, B.; Auffret, S.; Dieny, B.: Extraordinary hall effect in thin magnetic films and its potential for sensors, memories and magnetic logic applications. J. Phys. D—Appl. Phys. **41**(13), 135001 (2008)
49. Sandhu, A.; Sanbonsugi, H.; Shibasaki, I.; Abe, M.; Handa, H.: High sensitivity InSb ultra-thin film micro-hall sensors for bioscreening applications. Jpn. J. Appl. Phys. Part 2—Lett. Expr. Lett. **43**(7A), L868–L870 (2004)
50. van Kouwen, L., Botman, A., Hagen, C.W.: Focused electron-beam-induced deposition of 3 nm dots in as Scanning electron microscope. Nano Lett. **9**(5), 2149–2152 (2009)
51. Serrano-Ramón, L.E.: Thesis, Universidad de Zaragoza (in progress)

Chapter 4
Ferromagnetic Iron Nanostructures Grown by Focused Electron Beam Induced Deposition

We performed a detailed study of the magnetotransport properties and the microstructure in iron microwires grown by focused electron beam induced deposition as a function of Fe content, which varies from 45 to 70 %. Magnetotransport mechanisms have been investigated in detail over a wide range of temperatures and applied magnetic fields. Tuning of the Fe content and microstructure in the microwires has allowed us to explore the metal–insulator transition. We have also observed a progressive evolution in the magnetoresistance, from anisotropic magnetoresistance behavior to intergranular tunneling magnetoresistance. Additionally, Hall effect measurements reveal a giant anomalous Hall effect contribution due to the strong scattering of electrons produced by the magnetic impurities, which are present as iron and iron oxides species embedded in an amorphous Fe:O:C matrix.

4.1 Introduction

As discussed in Chap. 3, thanks to fabrication techniques such as FEBID and FIBID [1, 2], it is possible to control the growth of nanomaterials in a single step on any type of substrate, even with precision at the nanometric scale of the area of interest, the lateral width and thickness.

As a second paradigm for the nanofabrication of magnetic patterns using these techniques, recently details of the growth of Fe-based magnetic nanostructures by FEBID using different-material precursors, such as $Fe(CO)_5$, $Fe_2(CO)_9$, and $Fe_2(CO)_{12}$ [3–10], have been published. Remarkably, it has been possible to grow pure Fe NWs using a UHV system [11–15]. However, none of these studies included magnetotransport measurements, studies of microstructure, and composition of Fe deposits grown by FEBID until the systematic study published by our group in collaboration with the Eindhoven University of Technology (TU/e) [16].

R. Córdoba Castillo, *Functional Nanostructures Fabricated by Focused Electron/Ion Beam Induced Deposition*, Springer Theses, DOI: 10.1007/978-3-319-02081-5_4,

4.2 Growth of Iron Microstructures by FEBID

The growth of Fe structures was performed by FEBID, in a commercial Nova 600i (FEI Company) "Dual Beam" at Eindhoven University of Technology (TU/e). The deposition parameters of the Fe deposits were as follows: precursor material, diiron nonacarbonyl, $Fe_2(CO)_9$; precursor material temperature, 28 °C; $T_{substrate} = 22$ °C ± 1 °C, dwell time DT = 1 μs, overlap OL = 50 %, refresh time RT = 0 s; $P_{base} = 2 \times 10^{-6}$ mbar; $P_{process} = 4 \times 10^{-6}$ mbar (precursor gas valve is fully open during the growth process), 2 kV $\leq V_e \leq$ 30 kV, 0.7 nA $\leq I_e \leq$ 22 nA. The precursor material needle is aligned about 150 μm from the surface of the substrate (*z*) and about 50 μm in the plane direction, that is, in *x*/*y*. The substrate is Si (p-type doping) that is subjected to an oxidation process to ensure good electrical insulation. With this method a 250 nm layer of insulating material (SiO_2) is obtained. These substrates were subjected to optical lithography processes to define four or six titanium or aluminum contacts for four-probe-geometry measurements. Typical Fe microwire dimensions for the transport measurements were 12 μm length (*l*), 500–1000 nm width (*w*), and 200 nm thickness (*t*).

The Fe deposits studied in this section were grown with various values of V_e and I_e to study the dependence of the deposits' composition on these two parameters. The Fe content analysis was carried out in situ by EDS, as described in detail in Sect. 2.3. The acceleration voltage of the electron beam selected in the analysis of deposits with dimensions 500 nm × 500 nm × 100 nm ($l \times w \times t$) was 2 kV, to avoid substrate contributions in the spectra analyzed. The only elements detected were Fe, C, and O, which correspond to the precursor material and potential contaminants from the process chamber. In Table 4.1, we summarize the results obtained for different Fe nanosquares grown with various V_e and I_e.

Table 4.1 EDS compositional analysis of Fe nanosquares grown at several FEB acceleration voltages (V_e) and electron currents (I_e). Adapted with permission from [16]. Copyright (2011), IOP

Sample	V_e (kV)	I_e (nA)	Fe (at. % ± 2 %)	C (at. % ± 2 %)	O (at. % ± 2 %)	Fe/O
1	2	1.61	52.2	23.6	24.2	2.2
2	2	6.70	72.1	15.8	12.1	6.0
3	2	21.6	78.4	13.4	8.20	9.6
4	5	0.71	60.9	15.0	24.1	2.5
5	5	2.00	66.8	13.5	19.7	3.4
6	5	9.90	73.6	13.7	12.7	5.8
7	5	15.5	75.2	16.5	8.30	9.1
8	5	22.1	76.3	15.9	7.80	9.8
9	15	4.30	69.6	12.5	17.9	3.9
10	15	17.6	74.5	13.1	12.4	6.0
11	15	22.3	76.2	12.3	11.5	6.6
12	30	4.20	59.0	17.7	23.2	2.5
13	30	17.0	70.1	16.5	13.4	5.2
14	30	22.0	71.5	15.7	12.8	5.6

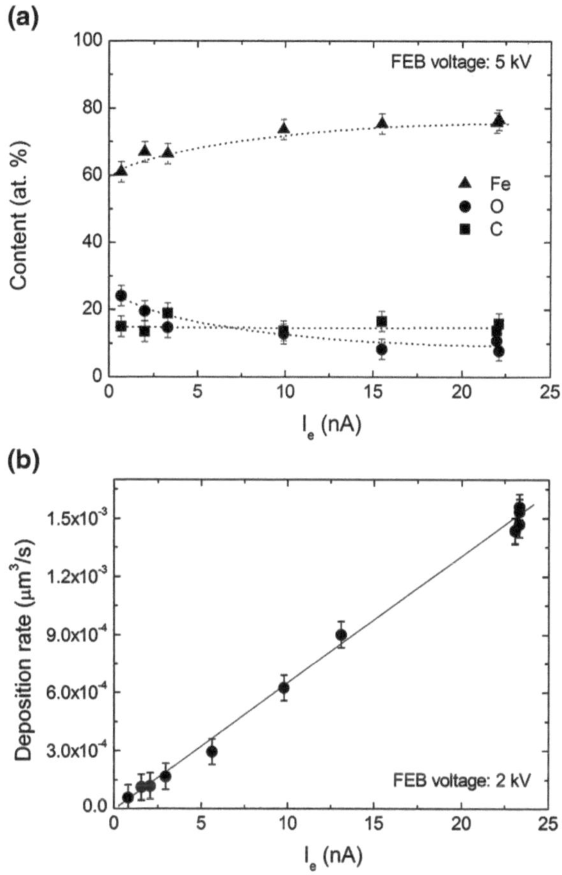

Fig. 4.1 **a** Composition of the deposits (at %) grown at 5 kV as a function of I_e. **b** Deposition rate of deposits grown at 2 kV as a function of I_e. Adapted with permission from [16]. Copyright (2011), IOP

As we noted for the Co deposits in Chap. 3, the maximum metal concentration in FEBID is obtained at high I_e (in the nanoamp range) and low V_e [17, 18]. Furthermore, we observe that the C content remains constant around $\sim 6 \pm 3$ (at %), while the Fe/O ratio varies. The maximum Fe concentration (~ 76 %) and the minimum O content (~ 8 %) are obtained at maximum I_e (22 nA).

In Fig. 4.1a, we show the results obtained in deposits grown at 5 kV. In the same way as described for Co nanodeposits in Chap. 3, this phenomenon may be related to a local heating effect arising from the use of high I_e [17–19]. Another effect that affects the Fe content is the high deposition rate that occurs at low V_e and I_e (Fig. 4.1b). A high rate would prevent the incorporation of contaminant species present in the process chamber in the final deposit [11, 20]. In order to determine the growth rate, we studied the growth rate as a function of I_e

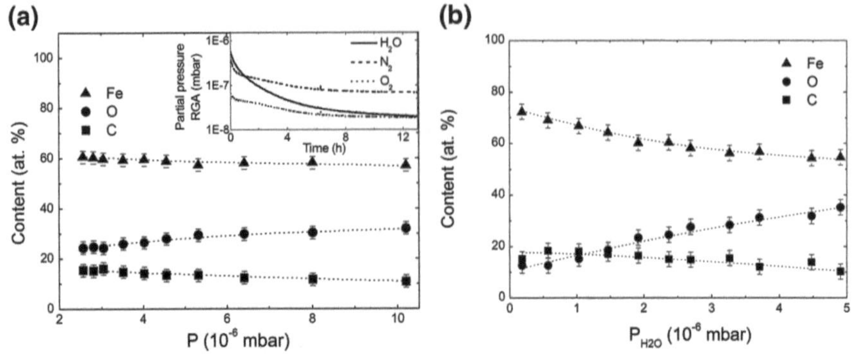

Fig. 4.2 **a** Fe, O, and C content (at %) of deposits grown at 2 kV and 13 nA as a function of the base pressure. The low chamber pressure in the experiment is taken after 3 h pumping. The inset shows the partial pressure of the main contaminant species as a function of the pumping time. **b** Fe, O, and C content (at %) of deposits grown at 5 kV and 22 nA as a function of the H_2O partial pressure. Adapted with permission from [16]. Copyright (2011), IOP

(Fig. 4.1b). The volume of the deposit was estimated by AFM. The linear dependence of the growth rate as a function of I_e indicates that the electron-limited regime applies and its slope is the volume per dose, 3.2×10^{-4} μm^3/nC, for a given value of V_e, in this case 2 kV.

Figure 4.2a shows the Fe content as a function of the base pressure of the process chamber. The base pressure varies with the pumping time, so these deposits were made from 0.5 to 3 h of pumping time. The inset shows the partial pressure as a function of the pumping time of the main contaminant species found in the process chamber measured by a residual gas analyzer (RGA) from MKS Instruments, model Vac-Check LM78RGA. The H_2O and O_2 partial pressure drops abruptly in the first 7 h of pumping. Weak dependence on the Fe content is observed with the base pressure. However, the O content increases and the C content decreases at high base pressure. This would indicate that H_2O and O_2 at high partial pressures oxidized Fe and C [21].

In order to modify the Fe content of the deposits, H_2O vapor was added by means of the material precursor magnesium sulfate heptahydrate $MgSO_4(H_2O)_7$. H_2O vapor is released into the process chamber by a needle of the same type as used for the rest of the precursor materials and its H_2O base partial pressure is monitored by RGA. The results are shown in Fig. 4.2b, which shows a clear decrease of the Fe content when the H_2O partial vapor pressure is increased. An increase of the O content and a small decrease of the C content are also apparent. This indicates that H_2O vapor has a direct effect on the Fe oxidation and less effect on the C oxidation, which forms volatile species that are easily pumped out the process chamber [21]. The partial decomposition of the precursor material versus inclusion and the reaction of gases in the chamber determine the final composition of the deposits.

4.3 Microstructure of Iron Deposits at the Nanoscale

The study of the microstructure of Fe deposits helps us to explain the structural and magnetic properties, which will be developed in the next sections.

The Fe microwires were grown at 5 kV and 22 nA on thermally oxidized Si substrates (250 nm SiO_2), whose dimensions were 12 μm × 1 μm × 200 nm ($l \times w \times t$). The compositional analysis (at %) of the three microwires on which we focus in the study is as follows: (A) Fe:40 O:59 C:1, (B) Fe:64 O:28 C:8, and (C) Fe:76 O:12 C:12. In the low Fe content microwire (A), H_2O vapor was added during the growth process with H_2O partial pressure of 1.52×10^{-6} mbar, while the microwires (B) and (C) did not have H_2O vapor added.

TEM samples for cross-sectional views were prepared by the conventional method described in Sect. 2.5.1 using a "Dual Beam" microscope, Helios 600 NanoLab from FEI Company. The study of the microstructure of Fe microwires was performed using a Titan^3 High-base microscope at 300 kV with C_s spherical aberration corrector.

Figure 4.3 displays the HRTEM images of experiments performed on three microwires (A), (B), and (C). Sample (A) is a good example for the study of the microstructure of samples to which H_2O vapor was added and thus to compare the samples without added H_2O vapor. In sample (A) we observed the formation of spinel-type crystals, whose dimensions are ~200 nm in length—the thickness of the deposit—and ~20 nm in width. The analysis of the images by FFT shows that most of the lattice stripes correspond to iron oxides with high oxygen content, such as Fe_2O_3 or Fe_3O_4. An example of this is illustrated in the inset of Fig. 4.3b, where the FFT of a grain corresponds to the structure of spinel-type iron oxides. While quantitative analysis in this type of sample is not feasible, it is evident that the presence of spinel-type iron oxide is predominant in sample (A). This is in agreement with the compositional analysis carried out in the microwire, in which a high oxygen content was detected, probably forming an iron compound. The presence of metallic Fe in the deposit cannot be excluded as many of its periodicities are close to those of iron oxides, making it very difficult to distinguish it from them, although its contribution in the deposit would always be residual. However, sample (B) has a quite different microstructure, mainly formed by an amorphous matrix, as deduced from the absence of reflections in Fig. 4.3d and a predominant diffuse scattering in the FFT. This amorphous matrix contains small crystalline particles of 1–2 nm size, which can be easily indexed as Fe-α. Finally, sample (C) exhibits a microstructure similar to sample (B), as seen in Fig. 4.3e and f. Again there is an amorphous matrix, although with a smaller amount of nanocrystals compared with sample (B).

These results allow us to conclude that in the presence of high H_2O partial pressure, the deposition mechanism is different, forming spinel-type FeO_x crystals. As discussed in the Sect. 4.2, added water can act as a catalyst, directly oxidizing Fe and forming FeO_x crystals as observed in sample (A). At low water vapor partial pressures, the deposition mechanism creates an amorphous film of Fe, C, and O, which are a few Fe-α crystalline grains.

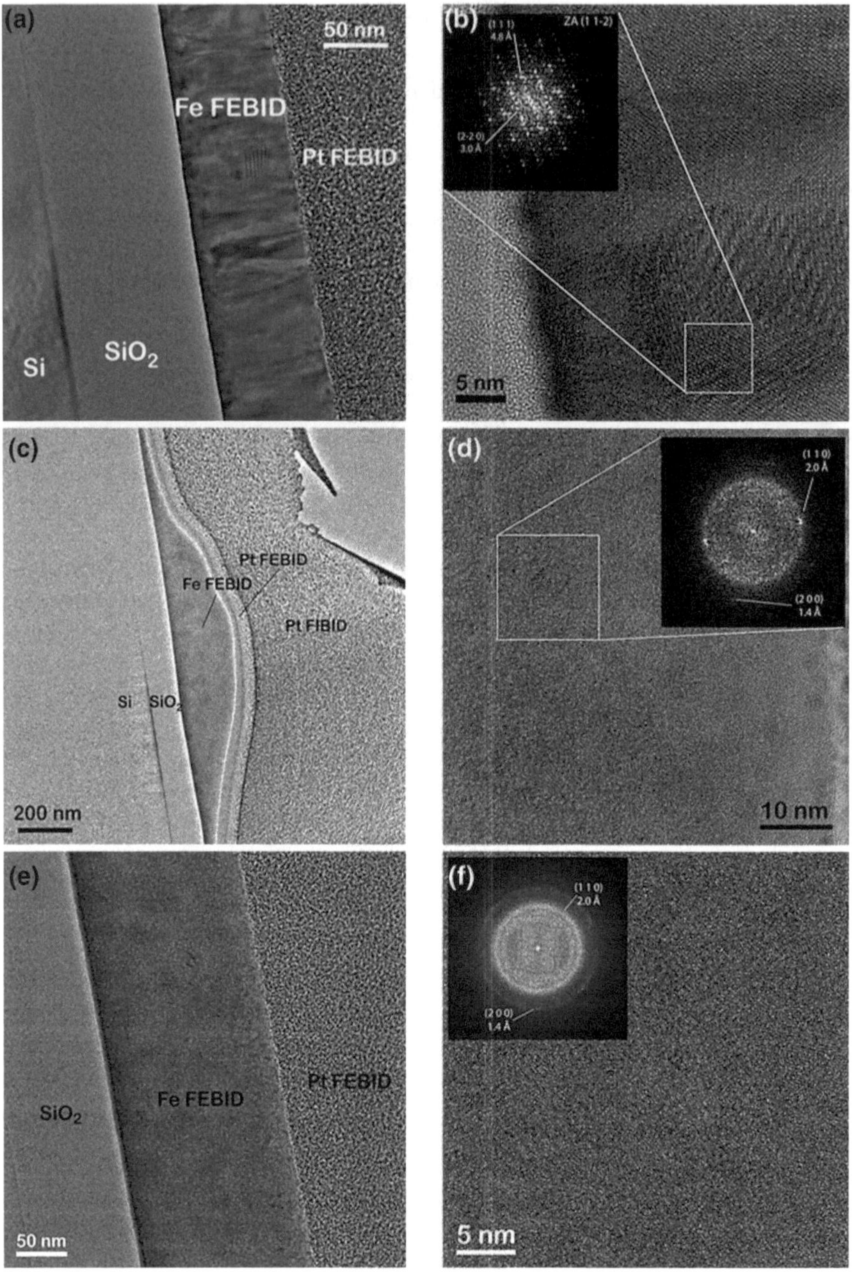

Fig. 4.3 HRTEM images at low and high magnification of the microwires (A), (B), and (C) with different Fe contents: **a**,**b** microwire (A), Fe:40 O:59 C:1; **c**,**d** microwire (B), Fe:64 O:28 C:8, and **e**,**f** microwire (C), Fe:76 O:12 C:12. The insets in **b** and **d** show the FFT of the areas indicated by white squares and the inset in **f** indicates the FFT of the entire image. Reprinted with permission from [16]. Copyright (2011), IOP

4.4 Electric Transport Mechanism

The electric transport mechanism in our deposits is related to its microstructure and far from trivial. The Fe microwires for this study were grown at 5 kV and 22 nA on thermally oxidized Si substrates (250 nm SiO_2) of dimensions 12 μm × 1 μm × 200 nm ($l \times w \times t$). The measurements were performed in the PPMS system described in Sect. 2.6.2 in four-probe geometry (see Fig. 4.4), working in a temperature range from 300 to 2 K.

Figure 4.5a shows the resistivity (ρ) as a function of the temperature of fifteen microwires with different composition (S1–S15), ρ at room temperature ranging from 186 $\mu\Omega$ cm (S1) to 3.84 × 10^5 $\mu\Omega$ cm (S15), depending on the Fe content. As expected from a Fe:O:C matrix, ρ depends strongly on the Fe content of the microwires. Thus, microwires with low Fe content exhibit a behavior typical of a semiconductor material; while the microwires having a high Fe content behave as a "dirty metal". In the case of S15, the resistivity increases to 1.25 × 10^7 $\mu\Omega$ cm at 130 K, the corresponding resistance (~7 × 10^6 Ω) being the maximum value that can be measured in the PPMS.

Figure 4.5b shows the residual resistivity ratio (RRR, defined as ρ_{300K}/ρ_{2K}). In S14 (Fe content ~56 %), ρ increases by more than one order of magnitude from 300 to 2 K (RRR ~ 0.05), whereas for the samples S1–S4 (Fe content close to 70 %) ρ hardly changes with temperature (RRR ~ 0.93).

The resistivity at 300 and 150 K is shown as a function of Fe content in Fig. 4.6, thus confirming a clear correlation between the Fe content and both ρ and T. As the Fe content decreases, the ρ increases; the slope of the curve changes drastically when the Fe content is about ~55 %. The difference bewteen ρ at 300 and 150 K becomes more than one order of magnitude when the Fe content is below 50 %. This behavior is similar to that previously observed by our group in Pt FIBID NWs [23], in which the metal–insulator transition (MIT) was observed as a function of Pt/C ratio.

The Fe content of the deposits decreases while the oxygen content increases, as seen in Figs. 4.1a and 4.2b, forming potentially insulating FeO_x crystals, which

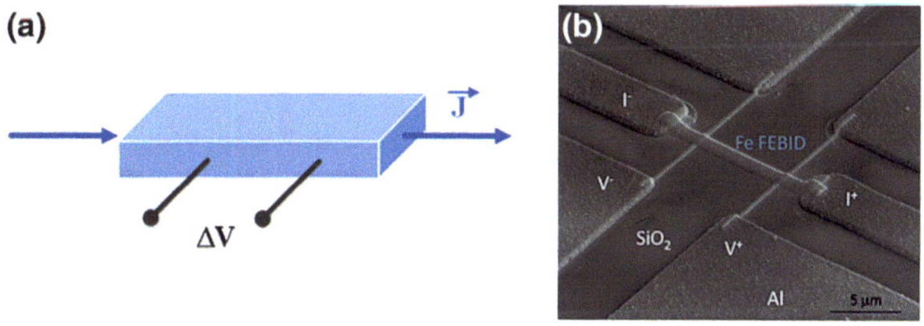

Fig. 4.4 **a** Diagram of the geometry used for the resistivity measurements by four-probe geometry. **b** SEM image of the measurement configuration

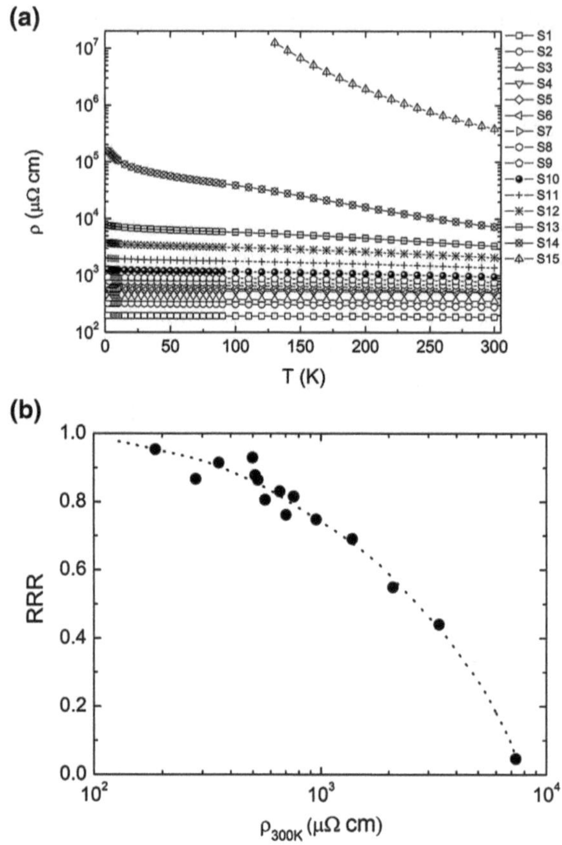

Fig. 4.5 **a** Resistivity of Fe microwires as a function of temperature. **b** RRR of Fe microwires as a function of resistivity at 300 K. Adapted with permission from [22]. Copyright (2012), IOP

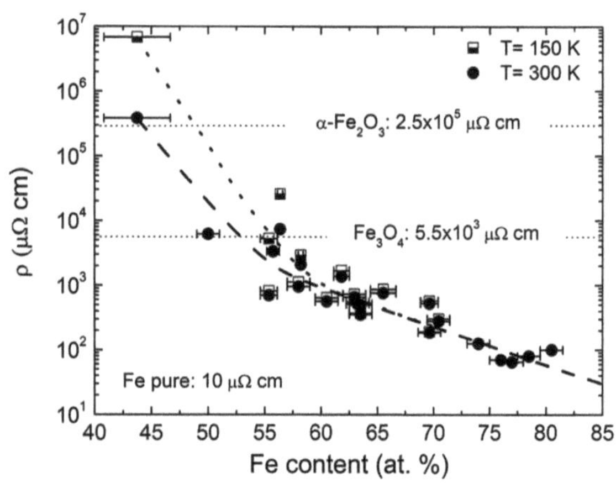

Fig. 4.6 Resistivity of Fe microwires at 300 and 150 K as a function of the Fe content. Resistivity values of Fe oxides at 300 K are indicated. Error bars correspond to the standard deviation of the spectra analyzed. Adapted with permission from [22]. Copyright (2012), IOP

increases the insulating character and the deposit's resistivity ($\rho_{Fe\ oxides} > 10^5$ μΩ cm). The abrupt increase in resistivity in samples whose Fe content is < 55 % would correspond to the formation of FeO_x crystals, as observed in the HRTEM images in Fig. 4.3a and b. In this range of Fe concentration, ρ at room temperature is higher than 10^3 μΩ cm, similar to Fe_3O_4, far away from the metallic conduction regime. In this case, the dominant contribution to the conduction mechanism arises from the FeO_x insulator crystals present in the microwire. This is the case for S14 and S15, the resistivities of which fit to an exponential dependence, $\rho(T) \sim \rho_0 \exp(T_0/T)^{1/2}$, indicating that the conduction mechanism can be explained by the "variable range hopping" (VRH) of electrons between localized states, which need to overcome a barrier or activation energy to move from one state to another.

For microwires with an Fe content >65 %, ρ is less than 550 μΩ cm (S1–S6), as expected in a system known as a "dirty metal". In this range of the Fe content, metallic conduction occurs in the amorphous and metal Fe:O:C matrix. Moreover, for Fe content from 55 to 65 % (S7–S13), the samples are thoroughly heterogeneous, a mixture of regions—sometimes dominated by the amorphous Fe:O:C matrix and sometimes by insulator iron oxides—having a combination of metallic and tunnel conduction mechanisms. The resistivity of samples with an Fe content up to 60 % is less than the Fe_2O_3 and Fe_3O_4 resistivities published in the literature [24, 25]. These results demonstrate that the conductivity becomes progressively degraded as the Fe content decreases, approaching a semi-insulating phase below 55 %. Thus, samples with a Fe concentration >65 % mark the limit of the metal percolation. The different Fe/O ratio and the microstructure found in the Fe-based microwires results in a MIT with a dramatic change in ρ dependence with temperature.

4.5 Magnetic Properties

4.5.1 *Magneto-Optic Kerr Effect and Longitudinal Magnetoresistance*

The best way to evaluate the potential of Fe-based microstructures grown by FEBID is to compare the magnetic behavior to that of pure Fe. Thus, pure Fe samples were fabricated, as a reference, with dimensions 12 μm × (0.5–1) μm × 30 nm ($l \times w \times t$), using optical lithography processes, EBL, and sputtering. The magnetic properties at room temperature were determined using the longitudinal magneto-optic Kerr effect (MOKE) technique at Eindhoven University of Technology (TU/e) in the framework of the collaboration already mentioned.

Figure 4.7 shows schematically the configurations for (a) longitudinal MOKE measurements and (b) longitudinal magnetoresistance (MR_{long}) measurements.

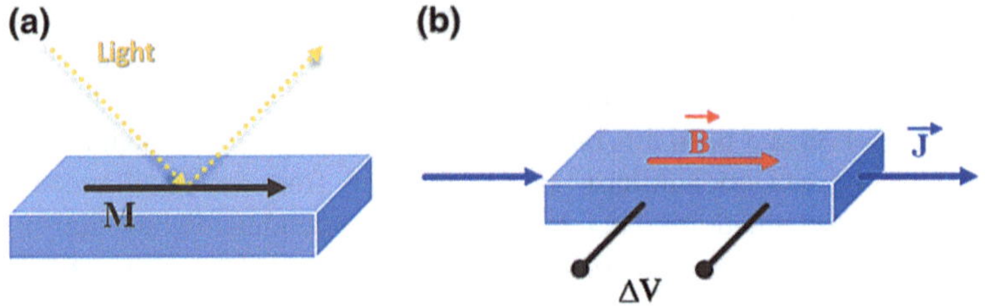

Fig. 4.7 **a** Diagram of the geometry used for MOKE longitudinal measurements. The magnetization is oriented parallel to the plane of the incident polarized light. **b** Diagram of the geometry used for MR_{long} measurements; the magnetic field is applied in the plane and direction of the current

Figure 4.8 compares MOKE and MR_{long} measurements as a function of applied magnetic field along the long axis of the structure for a pure Fe microwire (a,b) and an Fe FEBID microwire with composition Fe:72 O:15 C:13 (c,d). In (a), a square MOKE hysteresis loop is observed, as expected in pure Fe structures with easy axis along the long axis of the structure, while (b) shows the longitudinal MR of the same structure defined as MR_{long} [%] $= 100 \times [\rho(H)-\rho_{min}]/\rho_{min}$. The MR is attributed to the anisotropic magnetoresistance (AMR), well known for transition metals [26]. AMR reflects the dependence of the resistance of the sample on the angle between the current and the applied magnetic field. In this case, if the two are parallel the resistance is higher, and it is smaller when there is an angle between them. At high fields, all magnetic domains are aligned with the current (along the long axis of the microwire) to give high resistance, whereas, when the magnetic field is reduced, the magnetic structure relaxes towards an energetically more favorable position (multi-domain state), in which the magnetization direction rotates away from the direction of the current, reducing its resistance. Eventually, when the field changes sign and increases, the magnetization direction returns to being aligned with the current but in the opposite direction, corresponding to the MOKE hysteresis loop measured in (a). The results analyzed in Fe microwires grown by FEBID are shown in Fig. 4.8c and d. Regarding MOKE measurement, the first event is that the signal-to-noise ratio is low. This can be attributed to the low Fe content, the relatively small width of the microwire ($\sim$0.5–1 μm) compared to the laser diameter (2 μm), and the deposit' rounded surface, which can decrease the efficiency of the Kerr signal. Even with these experimental limitations, a clear hysteresis loop with a gradual change, indicating a less well-defined easy axis compared with the structure of pure Fe, is observed. The MR_{long} (d) shows a behavior similar to that of a pure Fe structure; it is a paradigm of the combination of gradual rotation of the magnetization and fast magnetization switching. Compared with the pure Fe microwire (b), small additional jumps in the signal corresponding to the movement of domain walls at different magnetic fields were observed. Furthermore, the relative decrease of MR_{long} in remanence compared with pure Fe could indicate the presence of a more complex magnetic domain structure during rotation of the magnetization. The first conclusion to be

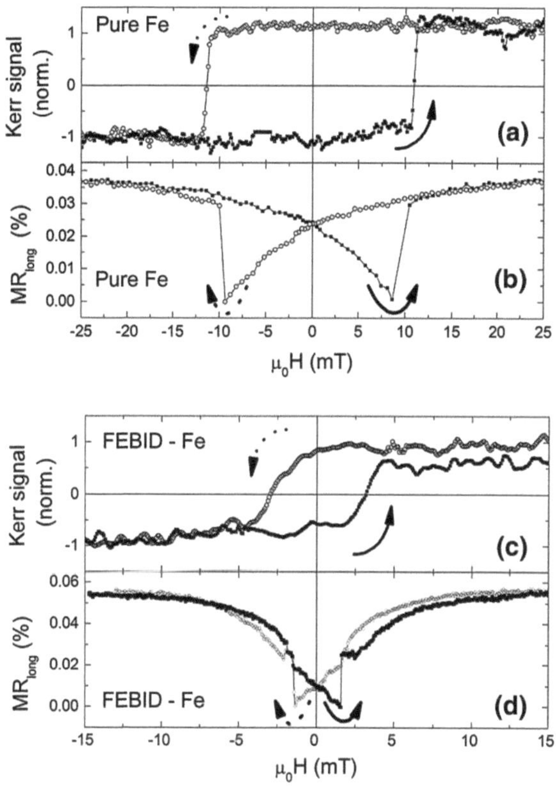

Fig. 4.8 MOKE measurements (**a**, **c**) and $MR_{longitudinal}$ (**b**, **d**) as a function of the applied magnetic field. **a**, **b** Pure Fe and **c**, **d** Fe microwires grown by FEBID with composition Fe:72 O:15 C:13. Arrows indicate the field sweep direction. Reprinted with permission from [16]. Copyright (2011), IOP

drawn from the MOKE measurements and the hysteresis loop observed in MR_{long} is that both indicate that Fe FEBID microwires have a genuine ferromagnetic behavior.

To investigate in detail the process of the magnetization rotation analyzed in Fig. 4.8d, wide-field Kerr microscopy was carried out in samples of Fe FEBID comparable to those used for their magnetic properties, but larger, 100 μm × 120 μm × 160 nm ($l \times w \times t$). This was necessary because the resolution of the technique was otherwise insufficient owing to the diffraction limit set by the wavelength used (∼400 nm).

In Fig. 4.9a, we show a hysteresis loop of this sample obtained from light intensity measurements in wide field Kerr microscopy images as a function of applied magnetic field. The insets in (a) indicate the Kerr images obtained when the magnetization of the sample is saturated in the direction of the applied field. The gray/black scale gives the magnetic contrast in the sample. In (b) and (c), we

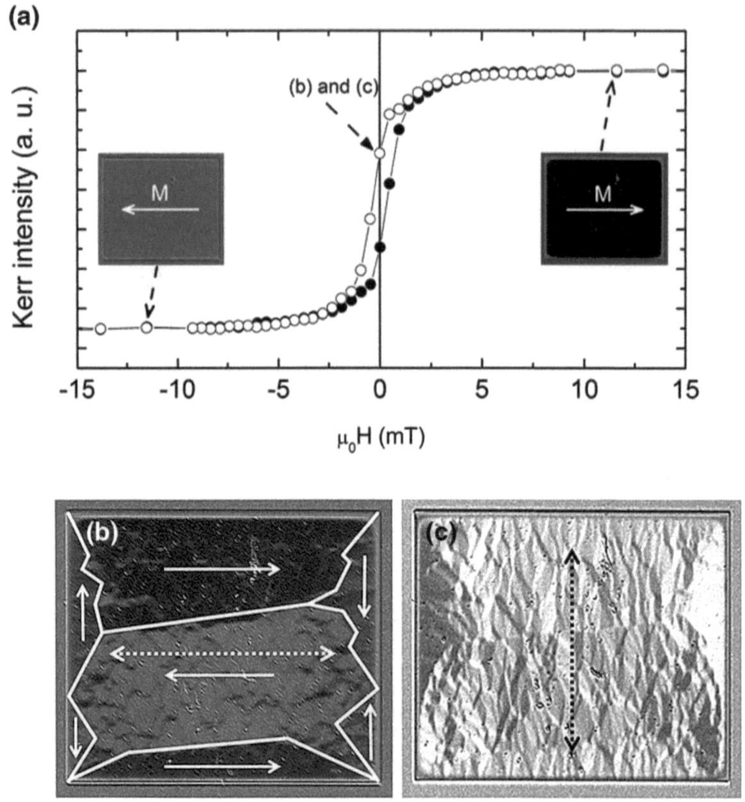

Fig. 4.9 **a** Hysteresis cycle obtained from the light intensity by wide field Kerr microscope images (longitudinal sensitivity) as a function of the applied field along the long axis of the structure. The insets show Kerr images taken when the magnetization of the film is saturated in the direction of the applied field; the gray/black scale indicates the magnetic contrast. **b** Longitudinal Kerr sensitivity image (dotted arrow) taken at the remanence state indicated in **a**. The flux closure domains are indicated by continuous arrows. **c** Image of transverse Kerr sensitivity (dotted arrow) at the remanence state indicated in **a**. Adapted with permission from [16]. Copyright (2011), IOP

can see the structure of domains in the remanence state shown in (a), for the longitudinal (b) and transverse (c) directions. The domain structure can be deduced from the contrast observed in (b) the longitudinal Kerr intensity and (c) the transverse Kerr intensity. Thus, we check the flux closure of the typical magnetic domains structure, confirming again the ferromagnetic behavior of the Fe structures grown by FEBID. Furthermore, we observe a complex structure of small domains at the film corners; probably due to slightly different thickness of the film at those ends or other defects. Therefore, the complex hysteresis loop and the low MR in the remanence state shown in Fig. 4.8d can be explained by the formation of many magnetic domains during the change of magnetization as shown in Fig. 4.9b.

4.5.2 Dependence of the Perpendicular MR on Temperature

In order to evaluate the magnetic behavior of the fifteen Fe microwires (discussed in Sect. 4.4) at different temperatures, we studied the dependence of the MR with the magnetic field applied perpendicular to the substrate (±9 T) calculated from Eq. (1.3).

In Fig. 4.10, we present a diagram and a SEM image of the geometry used for the $MR_{\perp}$ measurements. Taking into account the microwires' dimensions, coherent magnetization rotation under applied magnetic field is assumed.

The $MR_{\perp}$ of the four most representative samples are plotted in Fig. 4.11. In (a), we indicate the $MR_{\perp}$ of S1, a microwire with low ρ. This type of microwire needs a minimum magnetic field so that all magnetic moments are aligned out of the plane. The minimum field depends on the microwire geometry and the saturation magnetization (demagnetization field). $MR_{\perp}$, combined with MR_{long} presented in Fig. 4.8d, clearly indicates the effect of the anisotropic magnetoresistance (AMR). $MR_{\perp}$ is constant at high fields (>2 T) because the magnetic moments reach the saturation magnetization in the direction perpendicular to the plane. At fields between ±2 T, the magnetic moments turn from saturation back into the plane and $MR_{\perp}$ has a quadratic dependence on the field, as expected from coherent rotation of the magnetization [27].

An increase in the negative value of AMR is observed when the temperature drops: AMR(300 K) = −0.59 % and AMR(10 K) = −0.91 %. The presence of pure AMR signal in $MR_{\perp}$ is strong evidence of Fe percolation in the amorphous matrix of microwire S1. However, the AMR is high compared to the value for pure Fe, −0.2 % [28]. This can be explained considering the microstructure of the NWs, composed of an amorphous (Fe:C:O) matrix with small crystals of α-Fe (1–2 nm in size) [16]. The Fe atoms, which occupy many positions in the surface/interface, are less likely to suffer the effects of orbital momentum annihilation produced by the electric field of the crystal lattice and lead to greater spin–orbit coupling. This increase in the spin–orbit coupling may be based on this high AMR in a disordered system. There was a sharp decline of AMR when we increased the

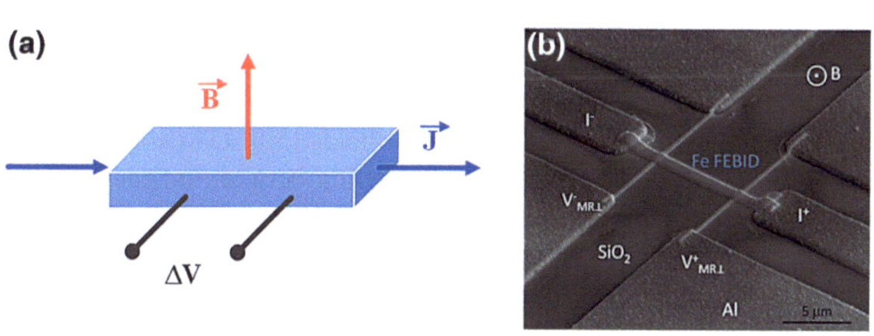

Fig. 4.10 **a** Diagram of the geometry used for $MR_{\perp}$ measurements, where the magnetic field is applied outside the plane and perpendicular to the structure. **b** SEM image of the configuration of $MR_{\perp}$ measurements

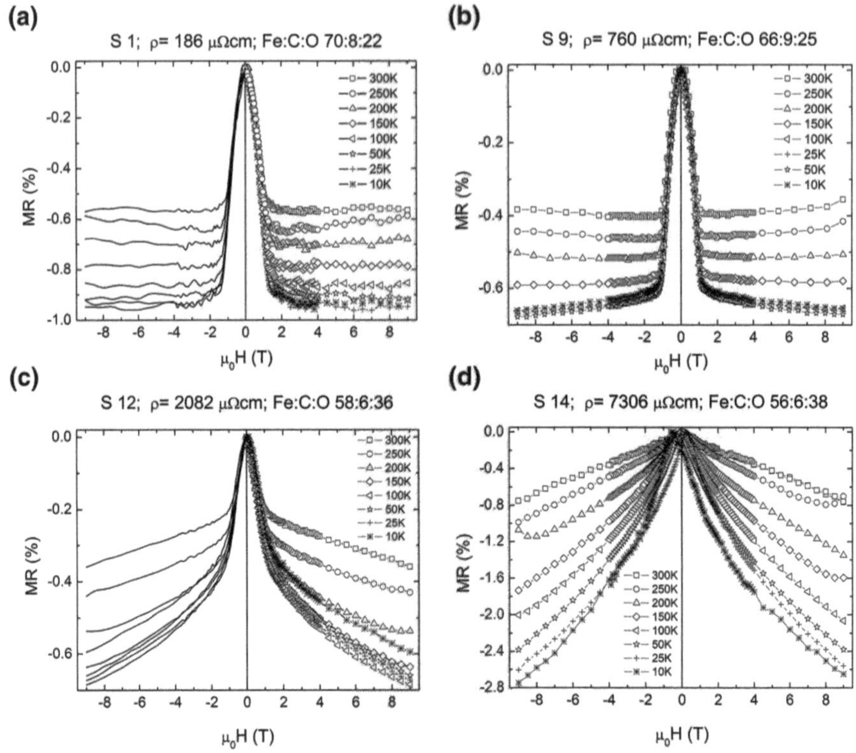

Fig. 4.11 $MR_{\perp}$ at different temperatures as a function of applied magnetic field for four representative samples. **a** S1, **b** S9, **c** S12, and **d** S14. S1 and S12 were measured exclusively at positive magnetic fields; hence we have drawn lines as a visual guide indicating the $MR_{\perp}$ extrapolated to negative values. Reprinted with permission from [22]. Copyright (2012), IOP

temperature from 10 to 300 K, which also indicates that the effects due to thermal fluctuations are high in systems with pronounced effects of magnetic atoms or interfaces with a defective bond [29].

$MR_{\perp}$ of sample S9 is shown in Fig. 4.11b, where both AMR and intergrain tunneling magnetoresistance (ITMR) effects are presented. The ITMR is due to the spin polarization of electrons and is produced by the transport that is established between the metal grains in granular ferromagnetic materials, which is sometimes difficult to saturate even under high magnetic fields [30]. ITMR is observed when the electric transport occurs by means of tunneling between magnetic grains. The probability of this happening depends on the magnetic orientation of the interfaces of these grains. $MR_{\perp}$ (±9 T) is −0.37 % at 300 K and −0.66 % at 100 K, becoming independent of temperature below 100 K. Thus, the $MR_{\perp}$ is dominated by AMR at high temperatures; from 300 to 200 K, $MR_{\perp}$ is completely saturated at 1.5 T. However, from 150 to 25 K, AMR and ITMR coexist. Below 150 K, $MR_{\perp}$ does not saturate at the maximum applied field, ±9 T, as observed in materials with $MR_{\perp}$ dominated by ITMR [29]. This fact correlates well with the microstructure in this composition range, consisting of Fe and Fe oxide grains embedded in an Fe:O:C matrix.

In more resistive samples (S12 and S13), a completely different behavior is observed (Fig. 4.11c and d, respectively). $MR_{\perp}$ does not saturate at the maximum field that can be applied in our system (±9 T). This behavior can be attributed to the ITMR associated with FeO_x granular systems [31].

In these microwires it is difficult to align the magnetization of grain interfaces with the external magnetic field and there is a linear slope at high field. An increase of $MR_{\perp}$ of S12 is obtained when the temperature decreases—$MR_{\perp}(300\text{ K}) = -0.36\ \%$ and $MR_{\perp}(100\text{ K}) = -0.69\ \%$—but below 100 K it remains constant. Assuming ITMR is the dominant effect in this microwire, ρ is close to the metal limit set by the Mott–Anderson theory for noncrystalline materials ($\rho = 3000\ \mu\Omega$ cm) [32], we concluded that S12 does not present a metallic conduction mechanism. A similar discussion can be applied to S14, in which the domain of the ITMR effect is even more evident, with MR = −2.7 % at 10 K and ±9 T. The ITMR effect dominates in these microwires owing to their increased O content (Fe/O = 1.61), which suggests the presence of a significant amount of semi-insulating FeO_x grains in the NW.

Figure 4.12a represents the $MR_{\perp}$ as a function of ρ_{300K} and the data corresponding to the four representative samples in Fig. 4.11 are marked by arrows. AMR and ITMR were determined at ±9 T. A clear dependence is observed, evolving from the $MR_{\perp}$ governed mainly by the AMR for microwires with moderate ρ, to high ITMR for high resistivity microwires. AMR is caused by the electron scattering due to spin–orbit coupling, while ITMR is caused by the tunnel conduction mechanism between the grains with spin polarization of electrons.

Finally, in Fig. 4.12b we show the $MR_{\perp}$ at ±9 T for the four samples on which the study was focused as a function of temperature. Thermal effects are more evident in samples in which the $MR_{\perp}$ is dominated by the ITMR. This is probably a direct consequence of the tunnel effect mechanism, in which the interfaces between grains are decisive [30]. In the case of samples in which $MR_{\perp}$ is dominated by AMR, the influence of temperature is less critical and may come from changes in phonon–electron scattering, combined with thermal fluctuations of the magnetic atoms in the amorphous matrix interface.

To summarize this section, the Fe-based microwires show a MIT transition defined by its Fe content and microstructure. Consequently, the $MR_{\perp}$ behavior has a gradual transition from AMR to ITMR as the Fe/O ratio decreases, which can be correlated with the microwire resistivity. The high value of $MR_{\perp}$ compared to pure Fe films [27] indicates that electron scattering is enhanced in Fe microwires grown by FEBID. As will be shown in the next section, this fact has an extraordinary impact on the Hall effect.

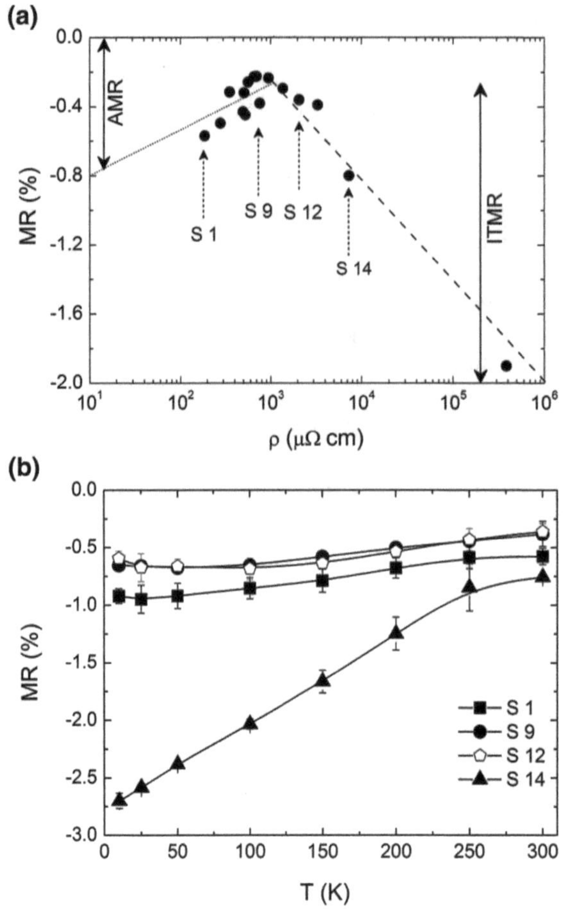

Fig. 4.12 **a** $MR_{\perp}$ of Fe microwires studied as a function of resistivity at room temperature; the data for the same microwires as in Fig. 4.11 are marked with arrows. **b** $MR_{\perp}$ of the same Fe microwires as in Fig. 4.11 as a function of temperature. AMR and ITMR were determined at ±9 T. Reprinted with permission from [22]. Copyright (2012), IOP

4.5.3 Study of the Giant Hall Effect

The study of the Hall effect in magnetic materials gives us useful information on the electron scattering processes, the carrier concentration, and the magnetization processes as a function of magnetic field. Hall resistivity is defined by the expression (1.4). A diagram and SEM image of the geometry used for the Hall effect measurements are given in Fig. 4.13.

In order to continue the analysis of the magnetic behavior of 15 Fe microwires at different temperatures, we performed a systematic study of the Hall effect over a wide range of temperature.

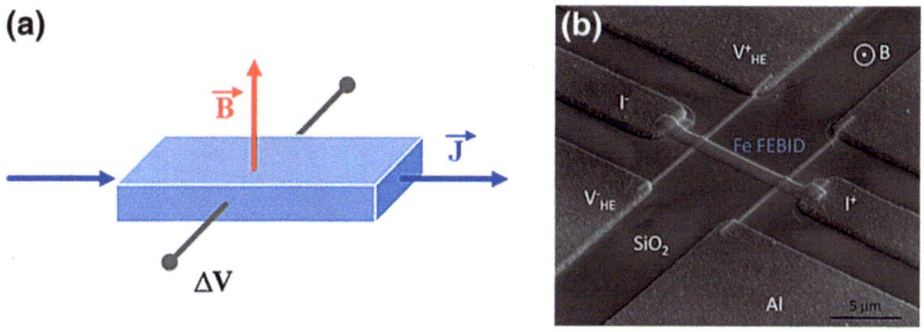

Fig. 4.13 **a** The Hall effect geometry used in the measurements. **b** SEM image of the measurement configuration

The regular contribution was determined by fitting the slope at high magnetic field, once magnetic saturation had been reached. The anomalous contribution at low magnetic field was determined by extrapolating the slope to $\mu_0 H = 0$. From the coincidence point for the linear fits of AHE and OHE, the NM_S anisotropy field is found; N is the perpendicular demagnetization factor for the microwire estimated to lie in the range [0.73–0.86] and M_S is the saturation magnetization [33], once magnetic saturation has been reached. In Fig. 4.14a, we see that ρ_H increases at low field until it reaches the saturation magnetization. The OHE and AHE fits are indicated by dotted lines in the isotherm at 10 K. M_S is extracted by this method and a clear dependence of M_S at 300 K as a function of longitudinal ρ is represented in Fig. 4.15. As expected, M_S decreases as a function of the longitudinal ρ from 1371 ± 16 to 706 ± 20 emu/cm^3. The error bars correspond to the standard deviations obtained in different fits of slopes. In comparison with pure Fe (M_S = 1700 emu/cm^3) and Fe oxides (M_S ~ 300–600 emu/cm^3) the magnetic moment per atom of Fe of these samples is similar to that of pure Fe, as M_S scales approximately with the Fe content (70 % of 1700 emu/cm^3 = 1190 emu/cm^3).

Previous results on the anomalous Hall effect in Fe films indicate that its magnitude correlates with the sample resistivity [34], as noted in our case.

However, the main information extracted from the Hall effect measurements comes from AHE analysis in terms of electron scattering. The first remarkable result is the high ρ_H in the three microwires (in the range of 10–150 µΩ cm) compared with values obtained in epitaxial Fe films [34], Fe microwires [15], and oxidized Fe films [28]. In fact, even in epitaxial Fe films with enhanced dispersion due to surface effects, ρ_H never exceeds 5 $\mu\Omega$ cm [34]. In oxidized Fe films, ρ_H ~ 1 $\mu\Omega$ cm [28]. Moreover, a large AHE at room temperature (ρ_H ~ 10–40 $\mu\Omega$ cm) has been observed in Fe_3O_4 epitaxial and thin films (thickness < 40 nm), an effect that was attributed to an increase of the electron scattering arising from antiphase boundaries formed during the growth of films on MgO(001) substrates [24]. Although AHE in Fe NWs should normally be small (<1 µΩ cm), additional mechanisms of electron scattering could lead to an AHE increase.

This hypothesis is certainly very interesting because the proper tailoring of the microstructure and composition of Fe NWs could result in a giant AHE (GAHE). The GAHE observed in Fe microwires fabricated by FEBID is due to the

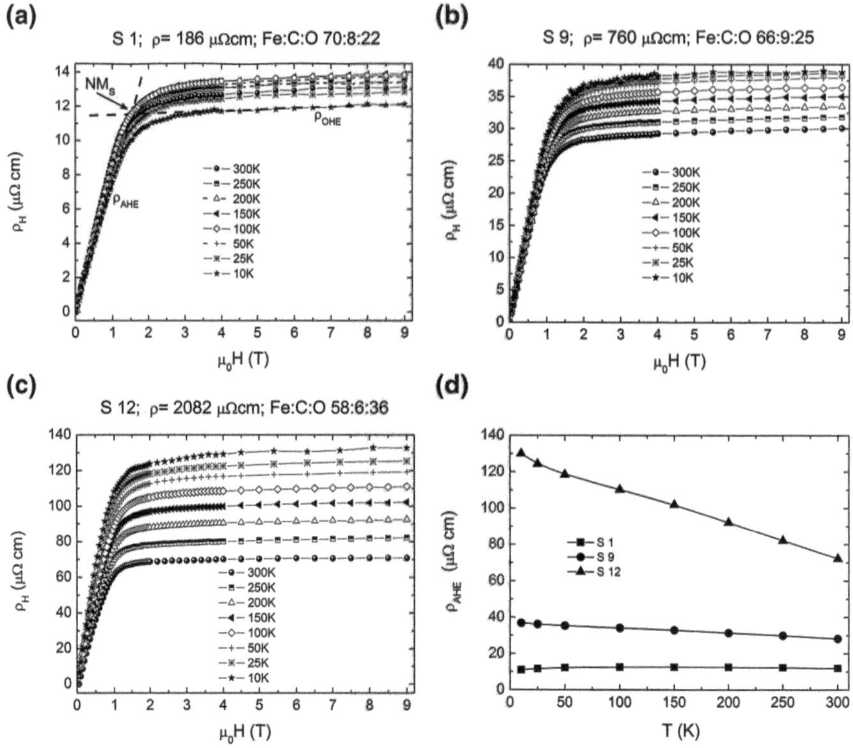

Fig. 4.14 Hall resistivity as a function of applied magnetic field at different temperatures. ρ_{OHE} is determined at high fields, whereas ρ_{AHE} is calculated by extrapolating the OHE slope at zero field. From the intersection of the linear fits the value of NM_S is obtained (N is the demagnetization factor, M_S saturation magnetization); an example is shown in (a) S1 at 10 K [18]. **a** S1; **b** S9; **c** S12; **d** anomalous Hall resistivity at 2 T as a function of temperature in the microwires shown in **a–c**. Reprinted with permission from [22]. Copyright (2012), IOP

heterogeneous and disorderly nature of the deposits. First, the microstructure enhances the electron scattering within the amorphous matrix, because the mean free path of the electrons will be strongly reduced in amorphous materials. Second, the presence of small grains of Fe and Fe oxides is another source of electron scattering in granular ferromagnetic materials [35]. It is well known that the AHE generally increases with longitudinal resistivity, on the assumption that the degree of ferromagnetism is maintained [36]. Thus, we argue that the explanation for the observed GAHE in our samples is provided by the microstructure, which produces enhanced electron scattering. Some of these electron scattering processes will be spin-dependent, contributing to GAHE.

The first key to the origin of the GAHE observed is its strong dependence on the longitudinal resistivity, as represented in Fig. 4.16a. In a first approximation, a linear dependence is observed, suggesting the skew scattering mechanism as a GAHE source. In this extrinsic mechanism, the electron conduction undergoes

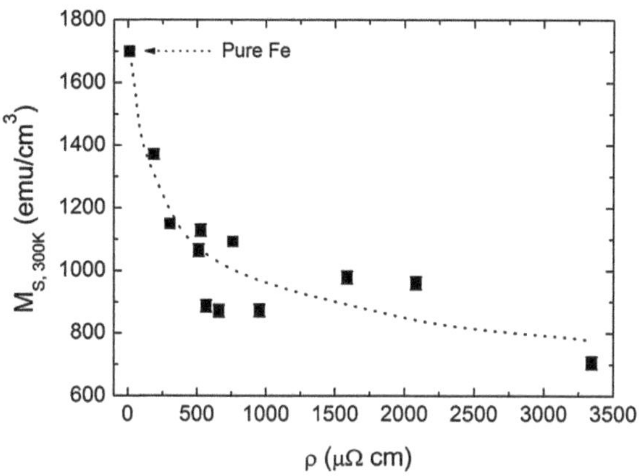

Fig. 4.15 Saturation magnetization at room temperature of Fe microwires based on their resistivity; evaluated through the analysis of the AHE measurements. Error bars were calculated from different fits of the slope of the curve ρ_H versus $\mu_0 H$. The value of the M_S (pure Fe) = 1700 emu/cm^3 is indicated in the graph. The dotted line is a guide to the eye that begins at the M_S of pure Fe. Reprinted with permission from [22]. Copyright (2012), IOP

spin-dependent scattering processes owing to spin–orbit interaction, modifying its trajectory at a certain angle, positive with respect to one spin direction and negative with respect to the other [37, 38]. In the case of Fe microwires, the scattering sources are magnetic impurities such as small Fe grains and Fe magnetic oxides with a certain distribution in the composition of the amorphous Fe:O:C matrix.

Owing to the heterogeneous nature of these Fe deposits, it is difficult to propose a detailed model for quantitative analysis of the data. However, Gerber et al. [35] found a way to model such heterogeneous systems assuming the longitudinal resistivity follows the Matthiesen rule, $\rho_{xx} = \rho_0 + \rho_S$, where ρ_0 contains sources of ballast scattering, which does not contribute to skew scattering, and ρ_S contains scattering processes that contribute to skew scattering. Thus, resistivity due to AHE can be expressed by

$$\rho_{AHE} = \gamma\rho_0\rho_S + \gamma\rho_S^2 \tag{4.1}$$

In this model, the skew scattering provides both the linear and square terms with respect to ρ_S. As a consequence, AHE variation from sample to sample and as a function of temperature will be mainly caused by changes in ρ_0 and ρ_S.

The selected microwires show a different behavior in this respect. S12, with high longitudinal resistivity, will have a higher contribution of ρ_0 than ρ_S. When this happens ($\rho_S \ll \rho_0$), Eq. (4.1) tends to $\rho_{AHE} \approx \gamma\rho_0\rho_S \propto \rho_{xx}$. A linear dependence between AHE and longitudinal resistivity (mainly due to the centers of ballast scattering) is expected, as seen in Fig. 4.16a. Furthermore, ρ_S does not vary significantly with temperature; ρ_0 is the term that governs AHE dependence on temperature. However, the quadratic term in expression (4.1) cannot be omitted in samples S1 and S9, in which the longitudinal resistivity is significantly lower and ρ_S is

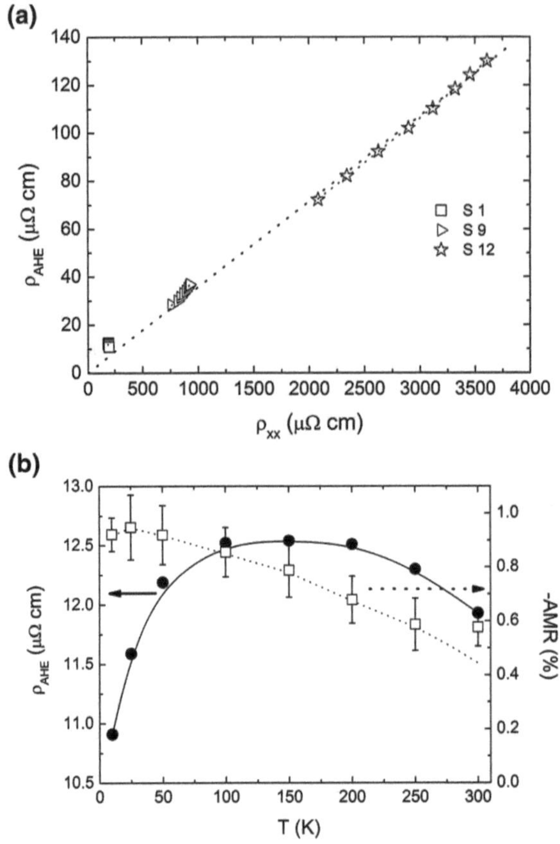

Fig. 4.16 **a** Anomalous Hall resistivity at 2 T as a function of the longitudinal resistivity for S1, S9, and S12. **b** Anomalous Hall resistivity at 2 T and anisotropic magnetoresistance at ± 9 T for S1 as a function of temperature. Adapted with permission from [22]. Copyright (2012), IOP

expected to make an important contribution to ρ_{xx}. This effect is illustrated by analysis of the results for S1, which has the lowest longitudinal resistivity. In Fig. 4.16b, ρ_H of this sample shows a maximum around 125 K. This can be explained by the relative contribution of the linear and quadratic terms in formula (4.1).

With increasing temperature ρ_0 increases as a result of an increase of the electron–phonon scattering. Thus, this explains the initial increase of ρ_{AHE} between 10 and 100 K, since ρ_S does not change significantly in this temperature range. However, ρ_S decreases with an increase in temperature because the magnetization, the spin polarization, and the magnetic interactions all decay with increasing temperature. This scenario is evidence for the dependence of ρ_S on temperature, which is confirmed by the AMR, also displayed in Fig. 4.16b, which reflects the same evolution of these parameters above 100 K. Thus, above 100 K, the quadratic term of Eq. (4.1) will decay significantly and can produce a full decrease of ρ_{AHE}. In Sect. 4.5.2, we mentioned that the decrease of AMR with temperature was higher than what one would expect in pure Fe. This shows that thermal fluctuations strongly affect the ferromagnetism in Fe microwires.

This effect is not surprising taking into account the sample's microstructure, which consists of a heterogeneous and amorphous Fe:O:C matrix with small Fe grains and Fe oxides. Nanometer crystalline particles of α-Fe and Fe oxide observed by HRTEM are prone to superparamagnetism at high temperature [39], thus decreasing the skew scattering process. The microstructure also suggests many magnetic atoms at the interface and defective bonds, which suppress magnetism and have a high tendency to be affected by temperature fluctuations.

4.6 Conclusions

Our results show that it is possible to produce functional microstructures using Fe FEBID that exhibit ferromagnetic character when the Fe/O ratio is high and the Fe content is higher than 65 at %. Adding water vapor during growth, we can control the Fe content, leading to an increase of FeO_x crystals in the deposit as seen in HRTEM images. Tuning the Fe/O ratio by adding water vapor, we could open a way to control the magnetic hysteresis loop of nanodeposits.

The resistivity of the deposits scales with the Fe content. Samples with high Fe content show a resistivity up to 19 times that of pure Fe and present a clear AMR. A transition from AMR to ITMR when the Fe content decreases (by increasing the oxygen content) is observed, indicating that the addition of water vapor during growth leads to the formation of FeO_x grains. The saturation magnetization obtained from Hall effect measurements indicated that the magnetic properties of amorphous Fe are quite similar to those of pure Fe.

The main result of this study has been the determination of a giant AHE, very high compared with that observed in pure Fe microwires, which has been explained in terms of a high skew scattering caused by the complex heterogeneous microstructure of these microwires. In microwires with high Fe content, a subtle interaction between the contributions of the spin-dependent and spin-independent scattering must be considered to understand the Hall resistivity dependence on temperature. This interaction is modulated by the important role played by thermal fluctuations in this heterogeneous system.

The use of the FEBID technique enables the growth of Fe microwires with totally different microstructure to that produced by other nanofabrication techniques, with a great impact on the magnetotransport properties, especially the observed GAHE, with future applications in new magnetic devices such as Hall effect sensors.

References

1. van Dorp, W.F., Hagen, C.W.: A critical literature review of focused electron beam induced deposition. J. Appl. Phys. **104**(8), 081301 (2008)
2. Utke, I., Hoffmann, P., Melngailis, J.: Gas-assisted focused electron beam and ion beam processing and fabrication. J. Vac. Sci. Technol. B **26**(4), 1197–1276 (2008)
3. Takeguchi, M., Shimojo, M., Furuya, K.: Fabrication of magnetic nanostructures using electron beam induced chemical vapour deposition. Nanotechnology **16**(8), 1321–1325 (2005)
4. Takeguchi, M., Shimojo, M., Furuya, K.: Nanostructure fabrication by electron-beam-induced deposition with metal carbonyl precursor and water vapor. Jpn. J. Appl. Phys., Part 1—Regul. Pap. Brief Commun. Rev. Pap. **46**(9B), 6183–6186, (2007)
5. Zhang, W., Shimojo, M., Takeguchi, M., Che, R-c, Furuya, K.: Generation mechanism and in situ growth behavior of alpha-iron nanocrystals by electron beam induced deposition. Adv. Eng. Mater. **8**(8), 711–714 (2006)
6. Takeguchi, M., Shimojo, M., Mitsuishi, K., Tanaka, M., Che, R., Furuya, K.: Fabrication of nanostructures with different iron concentration by electron beam induced deposition with a mixture gas of iron carbonyl and ferrocene, and their magnetic properties. J. Mater. Sci. **41**(14), 4532–4536 (2006)
7. Bruk, M.A., Zhikharev, E.N., Grigor'ev, E.I., Spirin, A.V., Kal'nov, V.A., Kardash, I.E.: Focused electron beam-induced deposition of iron- and carbon-containing nanostructures from triiron dodecacarbonyl vapor. High Energy Chem. **39**(2), 65–68 (2005)
8. Kunz, R.R., Mayer, T.M.: Catalytic growth-rate enhancement of electron-beam deposited iron films. Appl. Phys. Lett. **50**(15), 962–964 (1987)
9. Shimojo, M., Takeguchi, M., Mitsuishi, K., Tanaka, M., Furuya, K.: Fabrication of iron oxide nanostructures by electron beam-induced deposition. 6th Pac. Rim Int. Conf. Adv. Mater. Process. (PRICM-6), Mater. Sci. Forum **561–565**(1–3), 1101–1104 (2007)
10. Hochleitner, G., Wanzenboeck, H.D., Bertagnolli, E.: Electron beam induced deposition of iron nanostructures. J. Vac. Sci. Technol., B **26**(3), 939–944 (2008)
11. Lukasczyk, T., Schirmer, M., Steinruck, H.P., Marbach, H.: Electron-beam-induced deposition in ultrahigh vacuum: lithographic fabrication of clean iron nanostructures. Small **4**(6), 841–846 (2008)
12. Lukasczyk, T., Schirmer, M., Steinrueck, H.-P., Marbach, H.: Generation of clean iron structures by electron-beam-induced deposition and selective catalytic decomposition of iron pentacarbonyl on Rh(110). Langmuir **25**(19), 11930–11939 (2009)
13. Walz, M–.M., Schirmer, M., Vollnhals, F., Lukasczyk, T., Steinrueck, H.-P., Marbach, H.: Electrons as "invisible ink": fabrication of nanostructures by local electron beam induced activation of SiO_x. Angew. Chem. Int. Ed. **49**(27), 4669–4673 (2010)
14. Walz, M–.M., Vollnhals, F., Schirmer, M., Steinrueck, H.-P., Marbach, H.: Generation of clean iron nanocrystals on an ultra-thin SiO_x film on Si(001). Phys. Chem. Chem. Phys. **13**(38), 17333–17338 (2011)
15. Porrati, F., Sachser, R., Walz, M.M., Vollnhals, F., Steinrueck, H.P., Marbach, H., Huth, M.: Magnetotransport properties of iron microwires fabricated by focused electron beam induced autocatalytic growth. J. Phys. D—Appl. Phys. **44**(42), 425001 (2011)
16. Lavrijsen, R., Córdoba, R., Schoenaker, F.J., Ellis, T.H., Barcones, B., Kohlhepp, J.T., Swagten, H.J.M., Koopmans, B., De Teresa, J.M., Magén, C., Ibarra, M.R., Trompenaars, P., Mulders, J.J.L.: Fe:O:C grown by focused-electron-beam-induced deposition: magnetic and electric properties. Nanotechnology **22**(2), 025302 (2011). doi:10.1088/0957-4484/22/2/025302
17. Utke, I., Bret, T., Laub, D., Buffat, P., Scandella, L., Hoffmann, P.: Thermal effects during focused electron beam induced deposition of nanocomposite magnetic-cobalt-containing tips. Microelectron. Eng. **73–74**, 553–558 (2004)

18. Fernández-Pacheco, A., De Teresa, J.M., Córdoba, R., Ibarra, M.R.: Magnetotransport properties of high-quality cobalt nanowires grown by focused-electron-beam-induced deposition. J Phys D: Appl Phys. **42**(5), 055005 (2009)
19. Botman, A., Mulders, J.J.L., Hagen, C.W.: Creating pure nanostructures from electron-beam-induced deposition using purification techniques: a technology perspective. Nanotechnology 20(37), 372001 (2009)
20. Bernau, L., Gabureac, M., Erni, R., Utke, I.: Tunable nanosynthesis of composite materials by electron-impact reaction. Angew. Chem. Int. Ed. **49**(47), 8880–8884 (2010)
21. Shimojo, M., Takeguchi, M., Mitsuishi, K., Tanaka, M., Furuya, K.: Mechanisms of crystalline iron oxide formation in electron beam-induced deposition. Jpn. J. Appl. Phys. Part 1—Reg. Pap. Brief Commun. Rev. Pap. **46**(9B), 6247–6249 (2007)
22. Córdoba, R., Lavrijsen, R., Fernández-Pacheco, A., Ibarra, M.R., Schoenaker, F., Ellis, T., Barcones-Campo, B., Kohlhepp, J.T., Swagten, H.J.M., Koopmans, B., Mulders, J.J.L., De Teresa, J.M.: Giant anomalous hall effect in Fe-based microwires grown by focused-electron-beam-induced deposition. J. Phys. D—Appl. Phys. **45**(3), 035001 (2012). doi:10.1088/0022-3727/45/3/035001
23. Fernández-Pacheco, A., De Teresa, J.M., Córdoba, R., Ibarra, M.R.: Metal-insulator transition in Pt-C nanowires grown by focused-ion-beam-induced deposition. Phys. Rev. B **79**(17), 174204 (2009)
24. Fernández-Pacheco, A., De Teresa, J.M., Orna, J., Morellón, L., Algarabel, P.A., Pardo, J.A., Ibarra, M.R.: Universal scaling of the anomalous Hall effect in Fe_3O_4 epitaxial thin films. Phys. Rev. B **77**(10), 100403 (2008)
25. Mei, Y., Zhou, Z.J., Luo, H.L.: Electrical-resistivity of rf-sputtered iron-oxide thin-films. J. Appl. Phys. **61**(8), 4388–4389 (1987)
26. McGuire, T.R., Potter, R.I.: Anisotropic magnetoresistance in ferromagnetic 3d alloys. IEEE Trans. Magn. **11**(4), 1018–1038 (1975)
27. Leven, B., Dumpich, G.: Resistance behavior and magnetization reversal analysis of individual Co nanowires. Phys. Rev. B **71**(6), 064411 (2005)
28. Stankiewicz, J., Jimenez-Villacorta, F., Prieto, C.: Magnetotransport properties of oxidized iron thin films. Phys. Rev. B **73**(1), 014429 (2006)
29. García-García, A., Vovk, A., Pardo, J.A., Strichovanec, P., Algarabel, P.A., Magén, C., De Teresa, J.M., Morellón, L., Ibarra, M.R.: Tunneling magnetoresistance in Fe/MgO granular multilayers. J. Appl. Phys. **107**(3), 033704 (2010)
30. Serrate, D., De Teresa, J.M., Algarabel, P.A., Ibarra, M.R., Galibert, J.: Intergrain magnetoresistance up to 50 T in the half-metallic $(Ba_{0.8}Sr_{0.2})_2FeMoO_6$ double perovskite: Spin-glass behavior of the grain boundary. Phys. Rev. B **71**(10), 104409 (2005)
31. Ziese, M.: Extrinsic magnetotransport phenomena in ferromagnetic oxides. Rep. Prog. Phys. **65**(2), 143–249 (2002)
32. Mott, N.F.: Metal-insulator transition. Rev. Mod. Phys. **40**(4), 677 (1968)
33. Aharoni, A.: Demagnetizing factors for rectangular ferromagnetic prisms. J. Appl. Phys. **83**(6), 3432–3434 (1998)
34. Sangiao, S., Morellon, L., Simon, G., Teresa, J.M.D., Pardo, J.A., Arbiol, J., Ibarra, M.R.: Anomalous Hall effect in Fe (001) epitaxial thin films over a wide range in conductivity. Phys. Rev. B (Condens. Matter Mater. Phys.) **79**(1), 014431 (2009)
35. Gerber, A., Milner, A., Finkler, A., Karpovski, M., Goldsmith, L., Tuaillon-Combes, J., Boisron, O., Melinon, P., Perez, A.: Correlation between the extraordinary Hall effect and resistivity. Phys. Rev. B **69**(22), 224403 (2004)
36. Nagaosa, N., Sinova, J., Onoda, S., MacDonald, A.H., Ong, N.P.: Anomalous Hall effect. Rev. Mod. Phys. **82**(2), 1539–1592 (2010)
37. Smit, J.: The spontaneous Hall effect in ferromagnetics.1. Physica **21**(11), 877–887 (1955)
38. Smit, J.: The spontaneous Hall effect in ferromagnetics-II. Physica **24**(1), 39–51 (1958)
39. García-García, A., Vovk, A., Pardo, J.A., Strichovanec, P., Magén, C., Snoeck, E., Algarabel, P.A., De Teresa, J.M., Morellón, L., Ibarra, M.R.: Magnetic properties of Fe/MgO granular multilayers prepared by pulsed laser deposition. J. Appl. Phys. **105**(6), 063–909 (2009)

Chapter 5
Superconducting Tungsten-Based Nanodeposits Grown by Focused Ion Beam Induced Deposition

In this chapter we present the fabrication and characterization of superconducting tungsten nanodeposits grown by focused ion beam induced deposition. First, we study the influence of some deposition parameters, such as ion beam voltage, ion beam current, and incidence angle of the focused ion beam, on the nanodeposits' composition. Second, we present the superconducting properties of tungsten deposits of varying width, from microwires to ultranarrow nanowires of 50 nm width, by means of magnetotransport measurements. Finally, we study the nonlocal voltage generated in tungsten nanowires by nonlocal magnetotransport measurements. The results allow us to link this kind of superconducting nanostructures to potential applications in nanotechnology.

5.1 Introduction

In this section we describe the mechanism that occurs in the growth of tungsten structures by FIBID, a technique described in detail in Sect. 1.3.5, using $W(CO)_6$ as a precursor material. Next, we review the origin of superconductivity in this material, mentioning some of the most relevant previous work, which deals with their deposition conditions, the electrical characterization, and microscopy and scanning tunneling spectroscopy.

5.1.1 Tungsten Deposits Grown by FIBID

The deposition of W deposits grown by FIBID occurs by two different mechanisms in two different regimes. First, the growth starts with W atom nucleation at completely isolated points on the substrate surface. The substrate points where the nucleation starts is not directly related to the FIB scanning. The deposition rate depends on many parameters, such as substrate nature and roughness, ion dose, FIB scanning strategy, substrate temperature, and precursor flux. The initial

R. Córdoba Castillo, *Functional Nanostructures Fabricated by Focused Electron/Ion Beam Induced Deposition*, Springer Theses, DOI: 10.1007/978-3-319-02081-5_5,

tungsten nanodeposits increase in size and thickness, retaining their position and shape during the deposition, and merge with each other to form a continuous tungsten film. Then the second deposition regime starts, which occurs on the grown tungsten surface. The W deposition rate on the SiO_2 surface is higher than the W deposition rate on the deposited W surface. After the continuous film growth, the deposition rate is linear with respect to the ion dose. The electrical resistivity of the W deposits is 250 $\mu\Omega$ cm at 300 K, and shows a current density up to 3.53 MA/cm^2. Therefore the deposits are considered very suitable for the fabrication of nanocontacts in NEMS and MEMS [1].

The W FIBID evolution on SiO_2 found in Ref. [1] could be explained by a theoretical study published recently [2]. The researchers developed a model based on calculations from density functional theory that simulates the $W(CO)_6$ molecular interactions on the SiO_2 surface. They employed two different surface models: first, a fully hydroxylated SiO_2 surface (FOH), which could be considered as a real surface without vacuum conditions and irradiation; and second, a partially hydroxylated SiO_2 surface (POH), which could be considered as a substrate irradiated by an electron/ion beam or at high temperatures. Their results showed that the number of $W(CO)_6$ molecules adsorbed on a SiO_2 POH surface is higher than on a SiO_2 FOH surface. In the first case, spontaneous dissociation was observed, in which the chemisorption of a $W(CO)_5$ fragment on the surface led to extraction of one of the volatile CO ligands from the precursor molecule. This fact is explained by changes in the electronic properties and the charge transfer between precursor molecules and substrate. However, in the second case, a weak interaction (physisorption) between precursor molecules and substrate surface was observed.

Muthukumar et al.'s simulations clarify some aspects of interactions between the $W(CO)_6$ molecule and a SiO_2 surface, resulting in an useful tool to understand the early stages of W FIBID deposits on SiO_2, where W nanodeposits grow as independent islands [2]. However, in FEBID/FIBID processes, we should take into account a great many other reactions excluded in this study. For instance, beam interactions with precursor molecules and the substrate. These interactions can produce heating effects and charge on the substrate, besides the emission of secondary electrons or ions and scattered ions, inducing a $W(CO)_6$ molecule activation or dissociation.

5.1.2 Superconductivity in Tungsten Deposits Grown by FIBID

Tungsten deposits grown by FIBID using tungsten hexacarbonyl, $W(CO)_6$, as a precursor material have aroused great interest in condensed matter physics owing to its superconducting properties at low temperatures. This discovery was made by Sadki and coworkers in 2004 [3], who showed some of superconducting parameters of the W films obtained by magnetotransport measurements, such as the superconducting transition temperature $T_c \sim 5.2$ K, the upper critical field $H_{c2} \sim 9.5$ T, and the critical current density $J_c \sim 0.15$ MA/cm^2.

Previous results in the literature regarding amorphous tungsten nanodeposits with isotropic disorder present T_c higher than in pure crystalline deposits ($T_c = 0.01$ K). In the literature, we can find results for amorphous W films grown by other techniques that give T_c close to the value obtained in W FIBID nanodeposits. For example, W amorphous films evaporated and doped with Re have T_c from 3.5 to 5 K [4]. Furthermore, W amorphous films grown by sputtering and doped with Si or Ge have T_c from 4.5 to 5 K [5]. Even, W amorphous films deposited by CVD and doped with C have $T_c \sim 4$ K [6]. These results suggest that the amorphous nature of W nanodeposits grown by FIBID, in which disorder increases the electron screening length, thus reducing the electronic mean free path, is mainly responsible for the increase in T_c. Osofsky et al. proposed a model that explains the origin of the enhanced superconductivity in disordered metal systems [7]. According to the prediction of the Bardeen–Cooper–Schrieffer (BCS) model [8].

$$k_B T_c \sim 1.13\hbar\omega_D \exp(-1/VN(E_F)) \tag{5.1}$$

(where k_B is the Boltzmann constant, ω_D the Debye cutoff frequency, V the electron–phonon interaction, and $N(E_f)$ the density of states at the Fermi level) an increase in T_c can be due either to an enhancement of the electron–phonon interaction or the density of states.

After the discovery by Sadki and colleagues [3], other researchers conducted several studies on the growth and electrical characterization of tungsten nanostructures grown by FIBID. In Table 5.1 we summarize the most noteworthy results published in the literature. Some features are common to all of them, for example, the precursor material [$W(CO)_6$], the Ga^+ ion beam voltage (30 kV) and the amorphous nature of the nanodeposit.

Superconducting W-based FIBID is a powerful one-step growth technique involved in a vast number of applications in Nanotechnology. We give here some recently published examples. Superconducting W deposits have been used in the modification and repair of superconducting Nb devices, such as a SQUID device to fabricate an integrated SQUID susceptometer [13]. These devices have a high experimental spin sensitivity of 10^4 Bohr magnetons/Hz$^{1/2}$ and can be operated over a wide frequency range (DC to 1 MHz) and down to very low temperatures

Table 5.1 Characteristics and superconducting properties of tungsten deposits reported in the literature

Reference	I_i (pA)	W:C:O:Ga (at %)	$\rho_{300\text{ K}}$ ($\mu\Omega$ cm)	T_c (K)	H_{c2} (T)	J_c (MA/cm^2)
Sadki [3]	98	40:40:0:20	200	5.2	9.5	0.15 (2 K)
Luxmoore [9]	1	51:37:0:12	~200	~5.5	–	0.3 (3 K)
Spoddig [10]	10	17:35:4:13	–	5.2	6	0.015 (0.9 K, 0.5 T)
Li [11]	1–1000	23:68:0:9–41:37:0:22	–	5.1–6.2	–	–
Li [12]	20	34:53:2:11	125–325	5.1–5.4	–	1.0–1.74 (4.2 K)

(13 mK) [14, 15]. Finally, the fabrication of ultranarrow contacts without resistance [16] and 3D structures [17] also offer a wide variety of perspectives in MEMS and NEMS devices.

5.1.3 Superconducting Properties of Tungsten Films Grown by FIBID at the Nanoscale

The excellent topographic and spectroscopic characteristics of tungsten deposits grown by FIBID make them ideal for studying the superconducting state density using scanning tunneling microscopy/spectroscopy, (STM/S). This collaborative work summarized here was carried out by the group of Prof. Sebastián Vieira, Dr. Hermann Suderow, and Dr. Isabel Guillamón in the Low Temperature Laboratory of Universidad Autónoma de Madrid [18–21].

Superconducting W films of dimensions 30 μm × 30 μm × 0.2 μm were grown using the following deposition paramenters: precursor material, tungsten hexacarbonyl, $W(CO)_6$; precursor material temperature 55 °C; ion beam voltage $V_i = 30$ kV; ion beam current $I_i = 3$ nA; $T_{substrate} = 22$ °C ± 1 °C; dwell time 0.2 μs; overlap 0 %; refresh time 0 s; $P_{base} = 1 \times 10^{-6}$ mbar; $P_{process} = 1 \times 10^{-5}$ mbar. Films were grown on Au–Pd layer with 100 nm thickness previously deposited by sputtering on a Si substrate.

The temperature dependence of the superconducting gap is shown in Fig. 5.1a, which fits perfectly the curve from BCS theory for s-wave superconductors. W-based films are extreme type-II superconductors with weak coupling and near the dirty limit. The superconducting parameters were calculated from the microwires' resistivity at 300 K (275 μΩ cm) and the derivative of the upper magnetic field $B_{c2}(T)$ at $T_c = 4.15$ K, the value being extracted from STS measurements: $-(dB_{c2}/dT)_{T=Tc} = 2$ T/K. Thus, the coherence length value is $\xi(0) = 6.25$ nm, the London penetration depth is $\lambda(0) = 850$ nm, and the Ginzburg–Landau parameter is $\kappa = \lambda(0)/1.63\xi(0) = 83$. Other amorphous superconducting thin films showed similiar parameters [22]. Besides, the thermodynamic critical field is $\mu_0 H_c(0) = 40$ mT, the upper critical field B_{c2} was observed at 6.4 T and 0.1 K (defined as the field where the superconductivity is completely suppressed), and the superconducting gap is $\Delta_0 = 0.66$ meV. Simultaneous measurements of topography and tunneling spectroscopy curves in the same place allow one to construct an image of the vortex lattice, in which the vortices are arranged in a hexagonal configuration. In Fig. 5.1b, we observe a clear and well-ordered Abrikosov vortex lattice in experiments carried out at 1.2 K and 2 T.

Let us consider the growth of extra-flat and homogeneous tungsten thin films by FIBID, as an ideal method for the fabrication of superconducting systems for subsequent studies of the superconducting density of states as a function of the temperature and applied magnetic field by STM/S. Likewise, we must pay special attention to the selection of deposition parameters and the substrate roughness, as

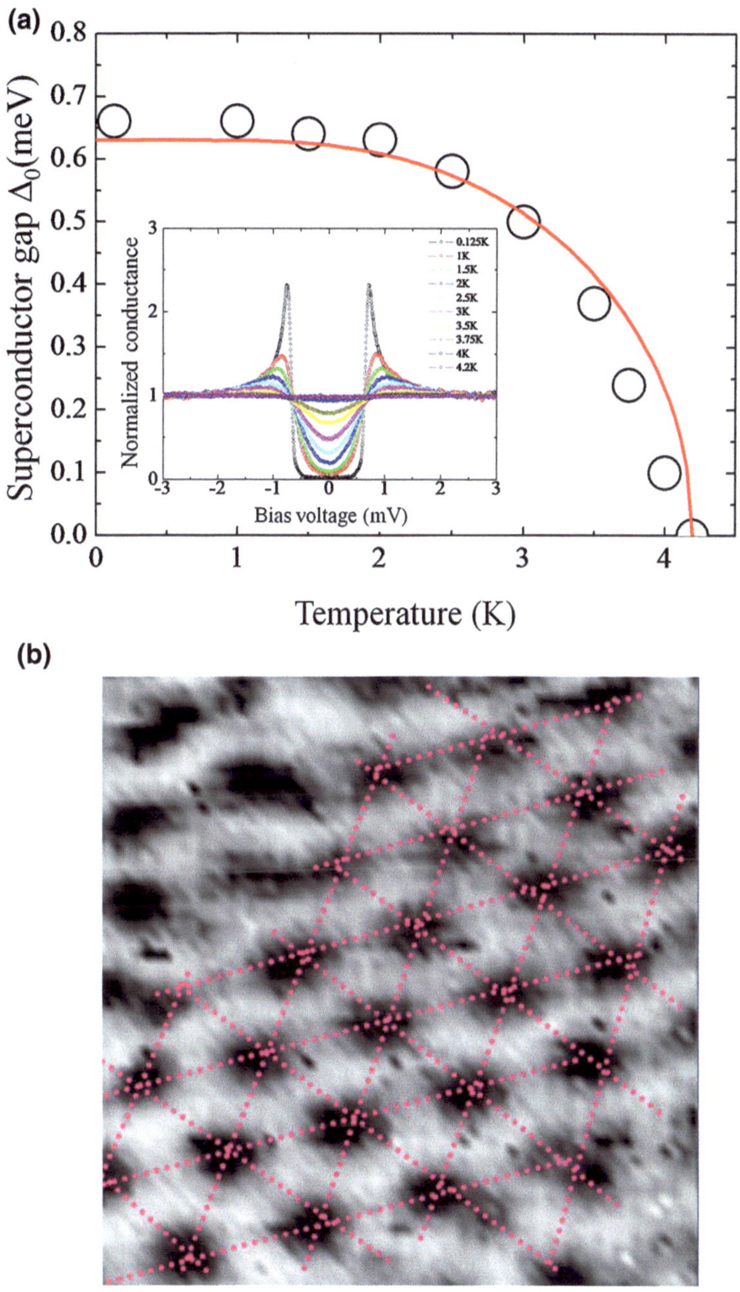

Fig. 5.1 **a** Superconductor gap as a function of temperature obtained from the fits to the tunneling conductance as a function of the voltage shown in the inset. *Red line* the fit from the BCS equation, taking $\Delta_0 = 1.76 k_B T_c$ and $T_c = 4.15$ K. Reprinted from [18]. **b** Hexagonal vortex lattice at 2 T and 1.2 K obtained from STS images constructed from the Fermi level conductance (the image shows a square of side 220 nm, where individual vortices where represented by *black areas*). The *pink dotted lines* show the well-ordered hexagonal vortex lattice [23]. Reprinted with permission from [20]. Copyright from (2009), NPG

the main factors for obtaining superconductor films with a low surface corrugation, below 1 nm.

In this chapter we report our investigation of W deposits grown by FIBID. First, we focus on the study of different deposition parameters, such as V_i, I_i, and the FIB incidence angle respect to the substrate, which can lead us towards to the control of the W content in deposits. Deposits were characterized by advanced techniques in spectroscopy and microscopy, such as EDS, AFM, and HRTEM. In addition, electrical measurements were carried out on microwires, nanowires, and ultra-narrow nanowires using the local and nonlocal measurements mode, which allowed the detailed study of the superconducting properties at low temperatures and at 0 T or under applied magnetic field.

5.2 Tungsten Structures Grown by FIBID

The fabrication of superconducting tungsten nanostructures was carried out by FIBID technique, explained in detail in Sect. 1.3.5 (Chap. 1), using a dual beam instrument, *Nova 200 NanoLab* or *Helios NanoLab 600* (FEI Company).

The deposition parameters are the same as those used in Sect. 5.1.3, except as follows: ion beam voltage (V_i) range 5–30 kV; ion beam current (I_i) range 1 pA–9.5 nA. In particular cases where we change any parameter this will be stated in the text. The precursor gas was injected into the process chamber using a needle inserted in the vicinity of the specimen, around 50 μm away in the *x*/*y*-direction and around 150 μm away in the *z*-direction of the substrate surface, to obtain an efficient FIBID process. A process of optical lithography using a lift-off method was performed to fabricate the Ti pads and connecting lines grown by e-beam evaporation with a thickness of 150 nm. A silicon wafer with 250 nm thermally grown oxide was used as a substrate.

The dimensions of the W-based structures, 40 μm × 2 μm × 1 μm ($l \times w \times t$) studied in this section were grown using 1 nA as I_i and modifying V_i to study its effect on the deposits' composition. The compositional analysis was performed in situ by EDS, avoiding exposure to enviromental conditions. Spectra were collected using a 10 kV electron beam voltage. Results obtained from the spectra are shown in Fig. 5.2. In Fig. 5.3a, we show a SEM image of a deposit grown at 30 kV.

The W and C content does not change more than 10 % (at %) as a function of V_i, and has a slight tendency to decrease with V_i. However, Ga content increases monotonically from 13 % ± 2 % at 5 kV to 23 % ± 2 % at 30 kV. This can be explained by the greater penetration depth of the Ga^+ ions at 30 kV than at 5 kV. The average of the ions' projected trajectories into the substrate is ~27 nm at 30 kV and ~8 nm at 5 kV (data obtained from TRIM simulations [24]). So, the implanted ion concentration will increase into the material because the probability of the ions escaping from the substrate decreases with increasing ion beam voltage. The O content remains constant around 3 % ± 2 %.

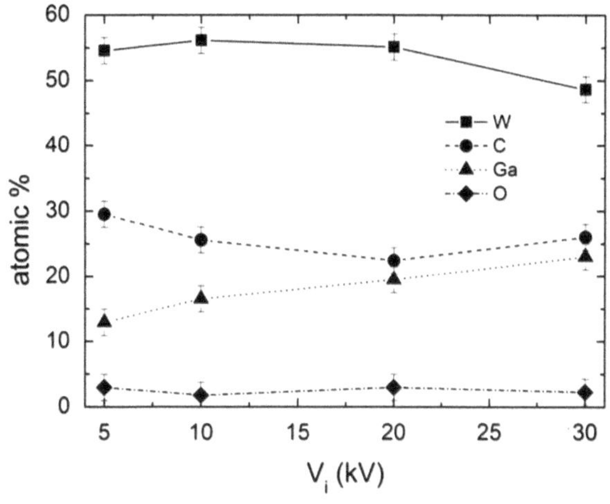

Fig. 5.2 Composition of W-based deposits as a function of the FIB voltage

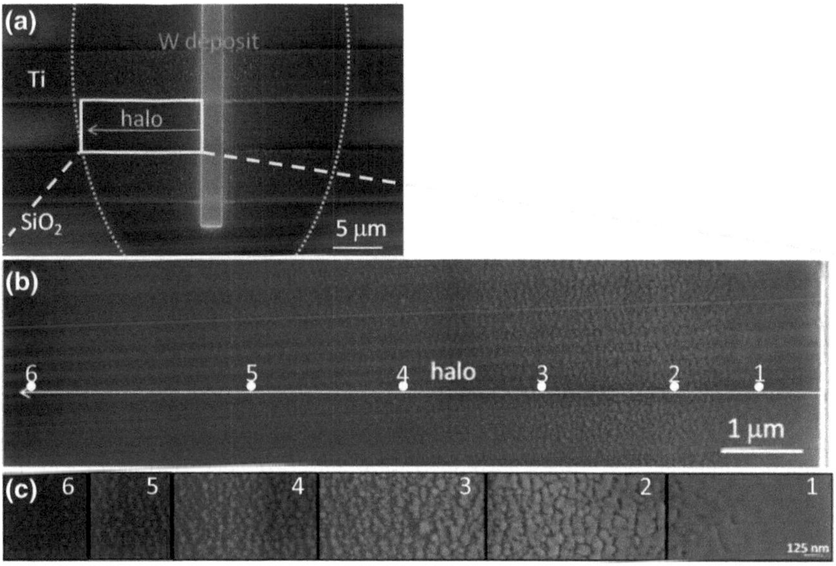

Fig. 5.3 **a** SEM image of the halo around the deposit; its extent is indicated by the *green dotted line*. **b** SEM image of the halo around the deposit from the area marked in (**a**) by the *light green rectangle*. **c** Higher magnification SEM images of the halo in (**b**) at the positions indicated by the numbers

We must not forget that around the W deposit there appears a secondary, undesired deposit with a few nanometers thickness, called a halo. The halo generated in W FIBID processes comes mainly from the decomposition of the precursor material produced by scattered Ga^+ ions and secondary electrons generated by them [1, 25] (see Fig. 5.4a), as we explained in Sect. 1.3.5. However, we cannot exclude the role

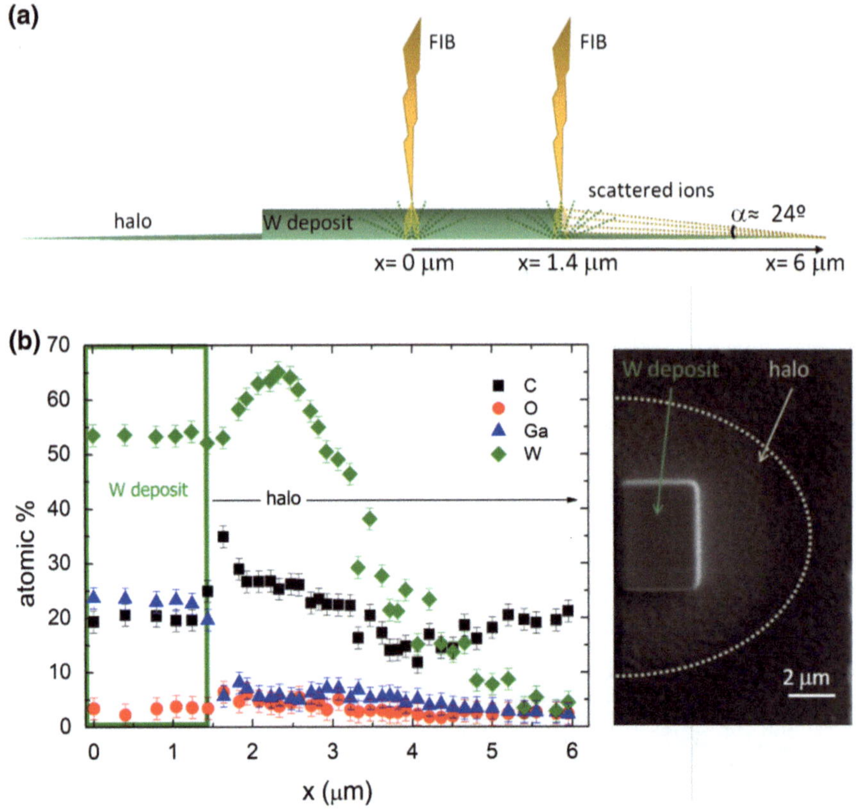

Fig. 5.4 **a** Drawing of the cross section of a W FIBID deposit, 4 μm × 3 μm in size. Note that the halo generated by the scattered Ga^+ ions and the *arrow* indicate the position in the plane from the SEM image in (**b**). **b** Relative composition of the W deposit represented in (**a**) and the halo as a function of the position in the plane. *Right-hand panel* SEM image of one of the W deposits studied

played by two additional effects, especially when we use beam currents in the nanoamp range, namely, material redeposition and a rounded profile instead of a Gaussian ion beam profile [26].

A halo is an unavoidable and unwanted effect that can extend for several micrometers in the plane, depending on the ion dose and the deposit thickness. In Fig. 5.3a, we can see the area where the halo grew. Obviously, the halo effect diminishes until it disappears far away from the W FIBID, as shown in Fig. 5.3b and c. In the area of the halo farthest away from the deposit, we observed isolated small grains 3 nm in size (see Fig. 5.3c). Size and grain density increases from left to right in the picture, until a continuous film is formed.

In order to study the halo composition generated around W FIBID deposits, some of them were grown at 30 kV and 0.92 nA, dimensions 3 μm × 4 μm × 2 μm ($l \times w \times t$), deposition time 5.5 min. Compositional analyses were performed in situ by EDS. Spectra were collected using a 10 kV electron beam voltage and results obtained from the spectra are shown in Fig. 5.4b. Substrate starts to be detected at $x = 2.58$ μm from the middle of the deposit, and its contribution increases as the distance in x increases. The substrate dependence is not represented in the graph to avoid confusion, but for that reason from $x = 2.58$ μm the sum of the C, O, Ga, and W percentages is not 100.

Deposit content (in atomic %) is W: 54 % ± 2 %, C: 19 % ± 2 %, Ga: 24 % ± 2 %, O: 3 % ± 2 %. However, the Ga content falls to 6 % ± 2 % at $x = 1.6$ μm, because the FIB only scans over the deposit, and the C content increases to 29 % ± 2 %, whereas W and O concentrations remains almost constant. From $x = 1.6$ μm to $x = 2.3$ μm, the relative W concentration increases to 65 % ± 2 %, which is the maximum concentration obtained, and then gradually decreases until the halo disappears. However, we must emphasize that the metal concentration (W + Ga) is always higher in the deposit than at any halo point. In addition, the fact that there is no direct FIB irradiation in the halo could explain the decrease in the amorphicity level and disorder in the halo. This fact could be related to a change in T_c in the halo. Later, we will show a study where deposits have been grown at different FIB angle in order to investigate the halo composition.

5.3 Microstructure of Tungsten Nanodeposits

The objective of this study is to analyze the microstructure of W FIBID deposits. Thus, the most appropriate technique is the acquisition and analysis of HRTEM images, which allow us to extract information about the crystalline structure at the atomic scale.

Tungsten deposits were grown on a commercial Cu grid with a carbon membrane (thickness ~10 nm). The thickness of the deposits should be less than 40 nm, so as to be able to obtain high-resolution images to collect the transmitted electrons that traverse the deposits. The deposition parameters were as indicated in Sect. 5.2.

In Fig. 5.5 we show an HRTEM image of a W deposit grown at 30 kV, 10 pA. The amorphous nature is revealed by the absence of diffraction rings in the fast Fourier transform (FFT) of the image. After this study, we can conclude that W deposits grown by FIBID consist of an amorphous matrix. Similar results were observed in W nanodeposits grown with a range of growth voltages and growth currents, 5–30 kV and 1 pA to 0.10 nA, respectively.

5.4 Influence of the FIB Incidence Angle

As mentioned previously, the W and Ga concentrations change significantly between the deposit and the halo around the deposit created by the partial decomposition of the precursor molecules by Ga^+ scattered ions. Therefore, the

Fig. 5.5 HRTEM image of W FIBID deposit grown on a standard Cu TEM grid at 30 kV and 10 pA with 20 nm thickness. *Inset* shows the absence of the diffraction rings in the FFT of the image

differences in composition combined with a lower level of amorphization of the halo could induce changes in the T_c.

In order to study the T_c variation as a function of the W content, we grew W FIBID deposits at different FIB incidence angle to tune the W content (see Fig. 5.6). The deposition parameters are $V_i = 30$ kV, $I_i = 10$ pA, time 4.17 min, dimensions: 1 μm × 1 μm × 0.2 μm ($l \times w \times t$), FIB incidence angle with respect to the substrate: 28°–90°, where at 90° the FIB is normal to the substrate. Compositional analysis was performed in situ by EDS; spectra were collected using $V_e = 5$ kV. Results (at %) are shown in Fig. 5.7.

In Fig. 5.6a, we show that the deposit is deformed in the y-direction; the deposition length in the y-direction increases from 1 μm to 2 μm when the FIB incidence angle is 38°, as shown in the sketch. This effect, which has already been studied by other researchers, is due to the dependence of the deposition length on the inverse cosine of the FIB incidence angle [27]. This deposit deformation in the y-direction decreases as the FIB angle with respect to the substrate approaches 90° as shown in Fig. 5.7b and c.

The W concentration increases from 56 % ± 2 % at 90° to 62 % ± 2 % at 38°, where we find the highest W content. Below 38°, W content decreases to 58 % ± 2 % at 28°. Although the differences in W composition are not really high, we observe a clear dependence with a peak at 38° FIB incidence angle. Ga concentration decreases monotonically from 17 % ± 2 % at 90° to 6 % ± 2 % at 28°. If we use a smaller FIB angle, a lower Ga^+ ion concentration that penetrates into the substrate is implanted in the material. The O content remains almost constant, 4 % ± 2 % from 90° to 38°, and then increases to 12 % ± 2 % at 28°, probably owing to the lower efficiency of precursor molecule

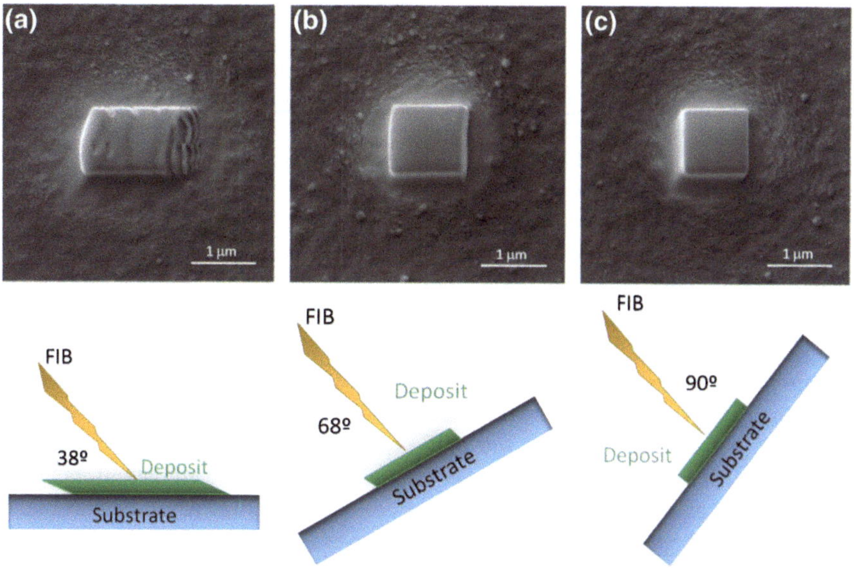

Fig. 5.6 *Top*: SEM images of W deposits grown with FIB incidence angles of (**a**) 38°, (**b**) 68°, and (**c**) 90° with respect to the substrate. *Bottom*: Each sketch shows the corresponding FIB incidence angle with respect to the substrate

Fig. 5.7 Composition of tungsten deposits as a function of the FIB incidence angle on the substrate. Note the cut in the *y*-axis

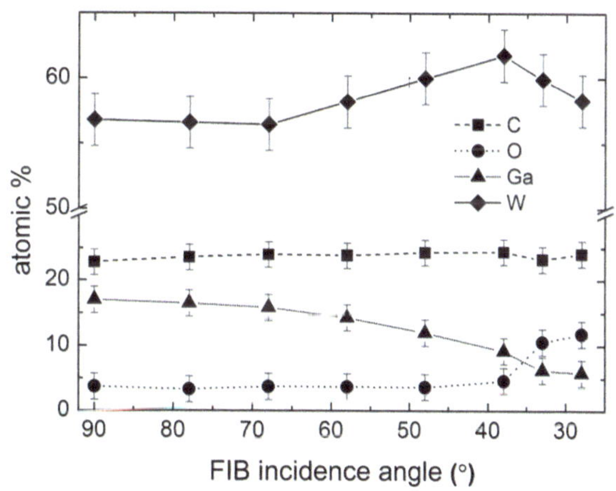

dissociation at that angle. The C content remains practically constant when the FIB angle is changed in the range studied.

SRIM software simulates the distribution range of projected ions by a Gaussian curve when we use low doses, although we note that the simple model simulated by the program is based on statistical distributions of ion collisions [24]. Table 5.2 indicates the average values of the distribution ranges on the substrate where the

Table 5.2 Data from TRIM simulations using Ga^+ FIB at 30 kV, and different FIB incidence angle on the SiO_2 (250 nm)/Si substrate

FIB incidence angle on the substrate	Longitudinal range (nm)	Projected lateral range (nm)	Projected radial range (nm)
At 90°	27	5	8
At 38°	17	21	22

ion concentration is maximal at 90° and 38° FIB incidence angle. The ion range has been described with respect to the final distribution of the ions as a function of the depth reached; the ion distribution exhibits azimuthal symmetry when a perpendicular FIB incidence angle is used on the sample. The lateral ion propagation can be summarized conveniently using an average, where y_i is the final value on the ion y-axis, z_i is the final value on the ion z-axis, and N is the number of ions calculated. Thus, the lateral range is just the end average of the ions' y–z displacement assuming a perpendicular FIB incidence angle. The projected lateral range is defined as the average of the absolute values of the projected lateral displacements on the x-axis. The projected radial range is the average of the range of x-axis radial displacement assuming a cylindrical symmetry. The results obtained in these simulations could explain why the Ga content decreases as a function of FIB angle.

After the composition study as a function of the incidence angle, we estimated the volume per dose by performing a cross section of each deposit. Thus, we can measure the deposit thickness and uniformity as a function of the growth direction (see Fig. 5.8). The standard method of the cross section process consists in: (1) Pt FEBID deposition on top of the sample to protect it and to increase the image contrast; (2) FIB milling from the top of the deposit; (3) SEM imaging of the cross section. The volume per dose (VD) indicates the efficiency of the growth process, in $\mu m^3/nC$.

In Fig. 5.8, we plot the VD of deposits as a function of the FIB incidence angle. This parameter is almost constant at 0.25 $\mu m^3/nC$ from 90° to 48°, but if we

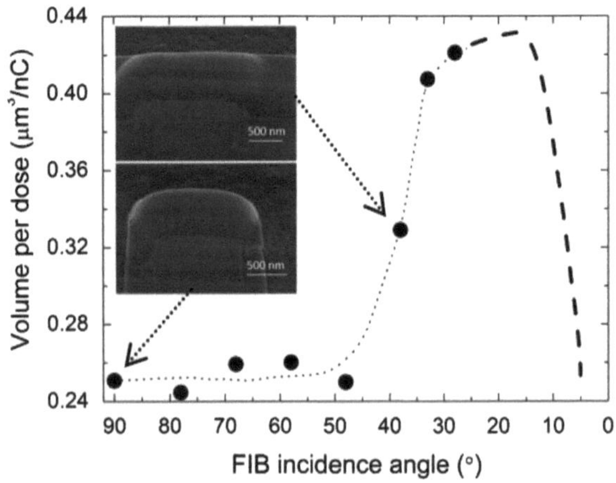

Fig. 5.8 Volume per dose of W deposits as a function of FIB incidence angle with respect to the substrate. *Insets* SEM images of cross sections of two W deposits, grown at 90° and 38°

decrease the incidence angle farther the VD increases to 0.42 μm^3/nC at 28°, which is the lowest angle we have studied. This remarkable increase in the growth rate of W deposits could be due to the higher ion concentration escaping from the substrate with high energy, which would be able to dissociate the adsorbed precursor molecules at a low FIB incidence angle. However, the VD dependence suggests that the VD continues to rise slightly until it descends abruptly when the FIB angle is excessively low with respect to the substrate surface (~15°), as in the FIB sputtering yield [26] or FEBID deposits [27]. Detailed analysis would be required to make this conclusion anything more than a qualitative explanation.

Once we had demonstrated changes in deposit content compared to the halo content, as well as deposits grown at different FIB incidence angles, we wanted to study the influence of these on T_c. For this reason, we studied the temperature dependence of the resistance of W nanostructures grown at different FIB angles, specifically those at which we observed changes in chemical composition (FIB angles of 38°, 48°, and 90°).

In Fig. 5.9 we show the normalized resistance of three W nanostructures as a function of the temperature. The critical transition temperature was determined as the midpoint of the transition (normalized resistance = 0.5) and thus we observe that the T_c difference between the three structures is no more than 0.1 K. T_c = 4.98 K at 38°, 5.08 K at 48°, and finally 5.04 K at 90°.

Thus, we do not find any evidence of significant changes in T_c of W structures deposited at different incidence angles (from 90° to 38°) and with different chemical composition.

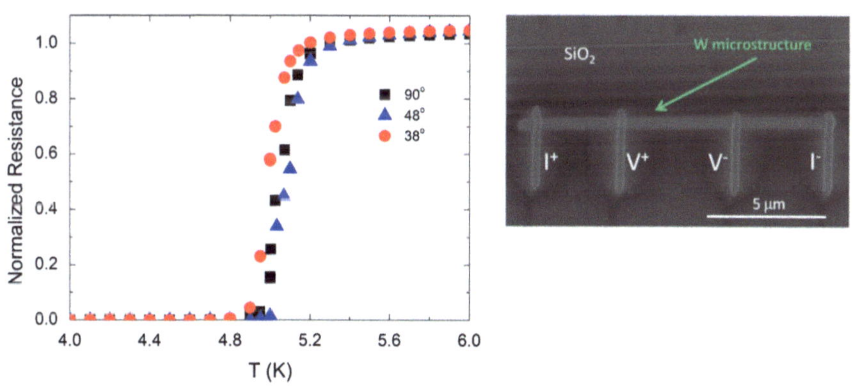

Fig. 5.9 Temperature dependence of the normalized resistance of three W nanostructures grown using different FIB incidence angles, 38°, 48°, and 90°. *Inset* SEM image of the W structure deposited at 38° FIB incidence angle

5.5 Local Electric Transport in Superconducting W-Based Deposits

In this section we study the superconducting properties of the W deposits with different lateral size, from microwires to ultranarrow nanowires, by four-probe electrical measurements with or without an external magnetic field perpendicular to the substrate.

5.5.1 *Superconducting W-Based Microwires*

The deposition parameters of W microwires grown on thermally oxidized Si (doped p-type) substrate are $V_i = 30$ kV, $I_i = 0.92$ nA, dimensions 39 μm × 2 μm × 0.3 μm ($l \times w \times t$). Ti pads used for four-probe measurements were fabricated by optical lithography techniques.

The electrical properties of W deposits were measured in situ in four-probe geometry to check that the electrical properties did not change with time. Resistance values measured ex situ in the PPMS even a year after the deposition do not vary by more than 10 % from the initial resistance measured in situ. In Fig. 5.10a we show a SEM image of one of these W microwires. In (b), we plot the resistance as a function of temperature; the critical transition temperature is $T_c = 4.95$ K (defined as the temperature at which $R_{normalized} = 0.5$).

In order to evaluate the applications of superconducting structures of this kind, another important parameter is the critical current, where the superconductivity disappears. The resistance dependence on the bias current at different temperatures (2.5–4.25 K) is shown in Fig. 5.11a. The critical current of microwires with a cross section of 0.6 μm^2 at 2.5 K is ~80 μA, which decreases as the temperature increases. Therefore, the critical current density at that temperature is about ~0.013 MA/cm^2. No significant changes were observed in the critical current density in such microstructures associated with composition variations or structural defects.

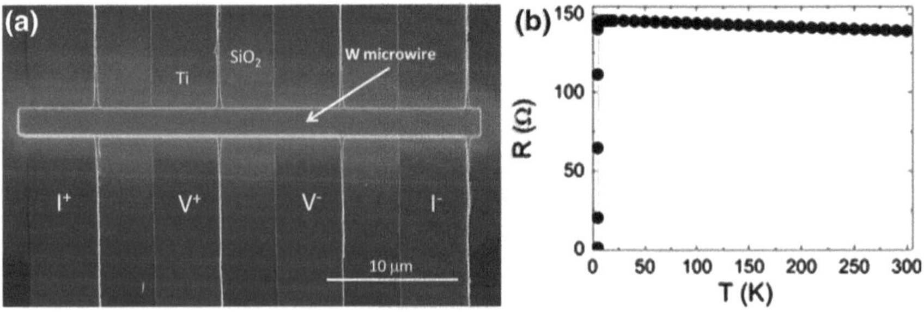

Fig. 5.10 **a** SEM image of a W microwire. **b** Resistance of the W microwire as a function of temperature

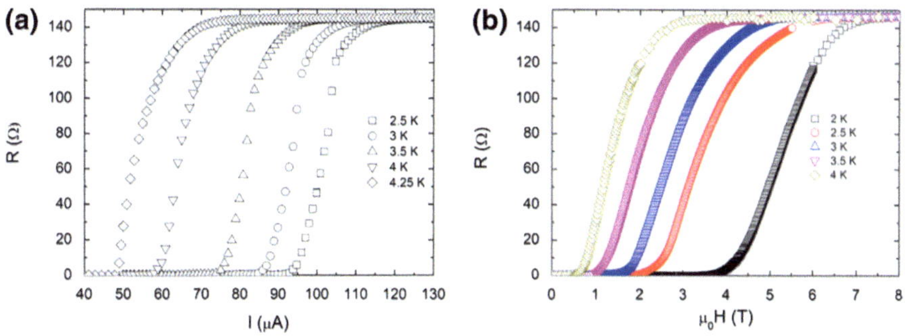

Fig. 5.11 **a** Resistance of the W microwire as a function of current at different temperatures (2.5–4.25 K). **b** Resistance of the W microwire as a function of the applied magnetic field at different temperatures (2–4 K)

Then we studied the microwire superconductivity as a function of an external magnetic field. We see how the field lines or vortices penetrate into the material, dissipating energy, which becomes the voltage measured by magnetotransport measurements. For a given temperature, the upper critical field B_{c2} is the parameter which indicates the suppression of superconductivity when the vortex density in the material is such that the distance between two vortices is less than the coherence length of the material at that temperature. We estimated B_{c2} from the resistance dependence study with applied magnetic field for various isotherms (2–4 K) that is represented in Fig. 5.11b. Thus, we know the range of fields for which the microstructures continue to exhibit superconducting behavior. For example, a finite resistance is not measured with the sensitivity of our measuring system at 2 K until the applied field is at least 3 T. From that value of magnetic field, the resistance continues to increase until superconductivity is completely inhibited at $B_{c2} \sim 7$ T. Obviously, B_{c2} decreases as temperature increases.

The superconducting parameters of W microwires studied in this section ($T_c = 4.95$ K, $J_c(2.5\text{ K}) \sim 0.013$ MA/cm^2, and $B_{c2}(2\text{ K}) \sim 7$ T) are of the same order as those obtained by other researchers [3, 9, 10, 12, 16, 28–30].

5.5.2 *Superconducting W-Based Nanowires*

In this section we deal with the growth and characterization of superconducting W NWs grown by FIBID with lateral size and thickness of ~200 nm. SEM images were performed in situ in the dual beam system, allowing high-resolution images of NWs for determining the lateral size. AFM measurements were conducted to investigate the thickness and uniformity along the length of the NWs. The W nanostructures were grown at $V_i = 30$ kV, $I_i = 29$ pA.

5.5.2.1 Electrical Measurements Geometry

Figure 5.12 shows a SEM image in which we can see the configuration of the W structures for electrical measurements by four probes in the local mode. First, we grew the W nanostructure, which consists of a vertical NW of 5 μm length and four NWs perpendicular to the first one of 3 μm length, used as contacts I^+, V^+, V^-, and I^-. The distance between the voltage contacts is 3 μm. Then, four Pt microwires FIBID were grown to connect the W nanostructure with the Ti contacts [31, 32]. The coincidence point between the FEB and FIB is perfectly aligned, so we adjusted the position where we want to grow the Pt microwires using the FEB, avoiding any FIB irradiation in the area where the perpendicular W NW was grown.

5.5.2.2 Temperature Dependence of the Resistance

In Fig. 5.13, we plot the resistance as a function of temperature for one of the W nanostructures studied. T_c ~5.1 K is determined as the midpoint of the transition ($R_{normalized} = 0.5$). The T_c obtained for W nanostructures by other researchers and even in our previous experiments is ~5 K [9, 10, 12, 16, 28–30]. In the inset of Fig. 5.13, we can see that the transition from the normal to the superconducting state does not shift when different currents are applied (0.1, 0.5, and 1 μA).

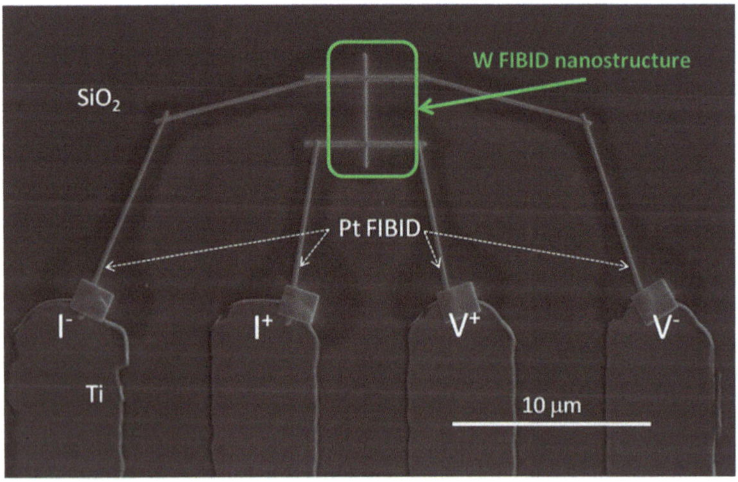

Fig. 5.12 SEM image of the local mode configuration for electrical measurements by four probes, the W nanostructure is indicated by the *green rectangle*, and the four Pt FIBID microwires that connect it with the Ti pads are indicated by *dotted white arrows*

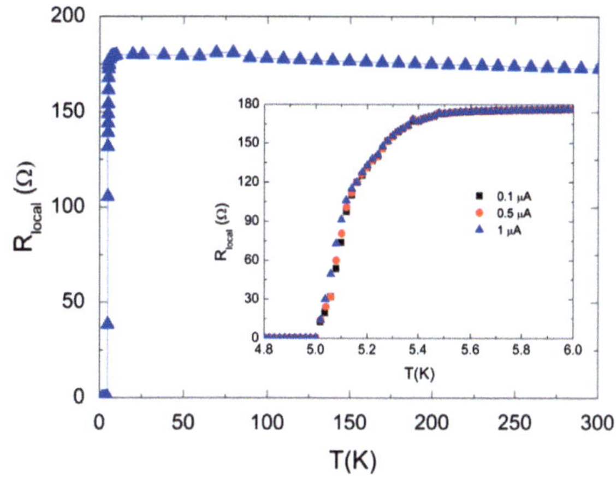

Fig. 5.13 Resistance versus temperature at different applied currents, 0.1, 0.5 and 1 μA. *Inset* In detail

5.5.2.3 Bias Current Dependence of the Resistance

First, we determined the critical current density of the superconductor NW without applied magnetic field at different temperatures (from 1 to 4.25 K). As an example, we show the normalized resistance as a function of the applied current at different temperatures without external magnetic field in Fig. 5.14. The critical current of the nanowire is ~37.5 μA at 1 K. This value decreases as the temperature increases and on approaching T_c. Therefore, the critical current density is ~0.094 MA/cm^2 at 1 K. Furthermore, no significant changes associated with small variations in composition or structural defects were observed in the critical current density in nanostructures of this kind.

Now, we focus on the evolution of energy dissipation in our superconducting material according to the applied current when an external magnetic field is applied. In this way, we show the most significant results obtained from local

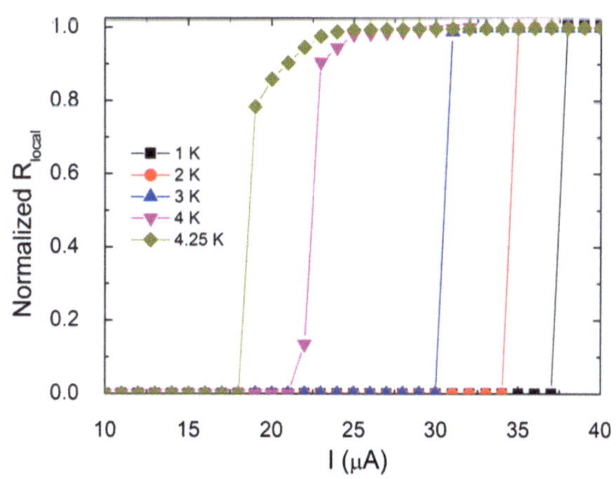

Fig. 5.14 Normalized resistance as a function of the applied current without external magnetic field at different temperatures

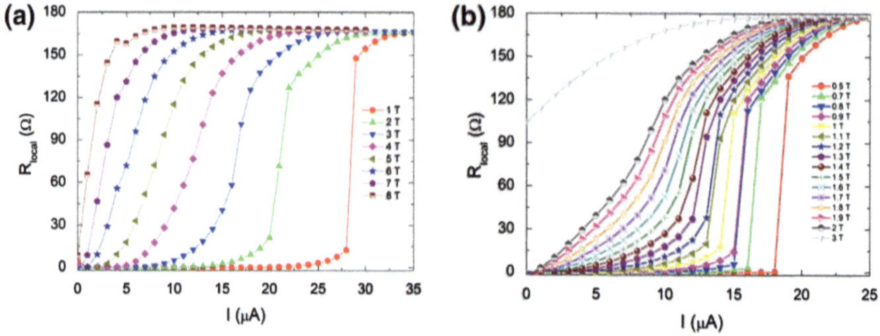

Fig. 5.15 Local resistance as a function of the bias current at different magnetic fields as indicated in the legend and at temperatures (**a**) 1 K and (**b**) 4 K

resistance curves as a function of the bias current at fixed temperature and at different magnetic fields.

Figure 5.15 shows the resistance as a function of the bias current at different magnetic fields and at (a) 1 K and (b) 4 K. First, we must emphasize that the energy dissipation due to the vortex motion caused by the bias current introduces a non-linear behavior that deviates from the dependence that it would follow if the vortex motion were governed only by the flux-flow regime, as explained in Sect. 1.6.2.1 (Chap. 1). The energy dissipation due to inelastic electron–phonon scattering would be the main cause of these instabilities of the flux-flow regime, which could generate a temperature increase of the electrons in the vortices [33, 34], which would create additional quasiparticles, thus decreasing the superconductor gap.

In Fig. 5.16 we represent the electric field ($E = \rho_f J$) as a function of the current density at low temperature, (a) 1 K, and (b) 3 K for fixed external magnetic field indicated in the legend, where ρ_f is the flux-flow resistivity and J the current density. However, for systems that are outside of the flux-flow regime we can apply the equation $J = 0.9\sigma_n E/b + J_c$ [35] (where σ_n is the conductivity in the

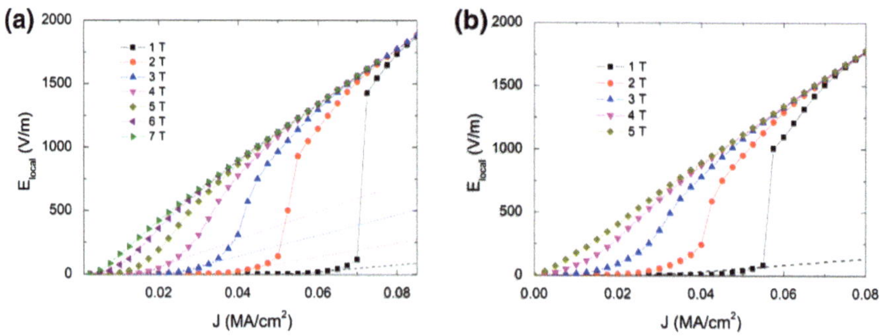

Fig. 5.16 Electric field as a function of the current density at different magnetic fields as indicated in the legend and at temperatures (**a**) 1 K and (**b**) 3 K

normal state, $b = B/B_{c2}$, and J_c is the critical current density that causes vortex depinning) for low values of applied magnetic field. In Fig. 5.16, we plot the partial adjustment of this equation at low current densities (dotted lines). Further analysis will be conducted in the future to determine what causes the flux-flow regime to arise out of the equilibrium state observed in our system.

5.5.2.4 External Magnetic Field Dependence of the Resistance

As introduced in Sect. 1.6 (Chap. 1), the upper critical field B_{c2} is the magnetic field at which a material's superconductivity is suppressed. Therefore, it is an important parameter to consider in our study. B_{c2} is estimated from the study of the dependence of resistance on the applied magnetic field for various isotherms (2–4.25 K). Thus, we can know the range of fields in which the structures continue to display superconducting behavior. A finite resistance is not observed at the lower value of bias current used (0.1 μA) at 2 K, with the sensitivity of our measurement system, until the applied magnetic field is at least 3 T. From this field, the resistance continues to increase until superconductivity is completely inhibited at $B_{c2} \sim 8.44$ T (defined as the field where $R = 0.9R_N$). As we increase the temperature, B_{c2} is reduced. In Fig. 5.17a we show the results obtained in the resistance versus applied magnetic field measurements at different applied currents and at 2 K. The results obtained in measurements of the resistance as a function of magnetic field at different temperatures (0.5–4.25 K) using 1 μA current are depicted in Fig. 5.17b.

From the curves of the resistance versus applied magnetic field for different applied currents, we extract B_{c2} values. In Fig. 5.18a we represent them as a function of applied current density in the range of temperatures studied and in Fig. 5.18b as a function of temperature at different bias currents. As we can see in Fig. 5.18a, B_{c2} hardly varies at low current density at a fixed temperature.

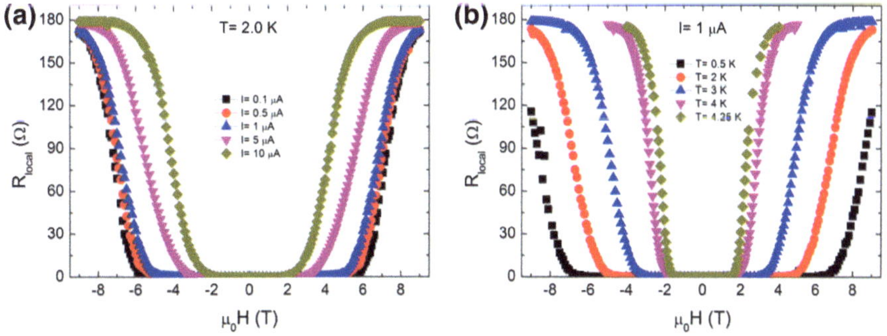

Fig. 5.17 **a** Local resistance as a function of applied magnetic field at 2 K and at bias currents of 0.1, 0.5, 1, 5 and 10 μA. **b** Local resistance depending on the applied magnetic field at different temperatures (0.5, 2, 3, 4 and 4.25 K) and at 1 μA bias current

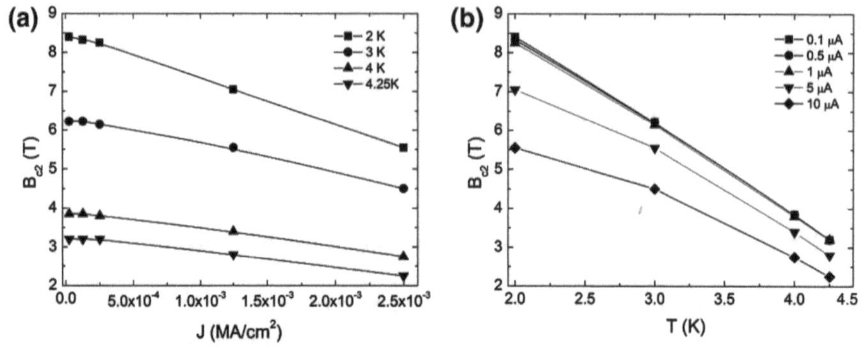

Fig. 5.18 Upper critical magnetic field (defined as the field where $R = 0.9R_N$) as a function of (**a**) the current density at 2, 3, 4, and 4.25 K and (**b**) the temperature at different bias currents (0.1, 0.5, 1, 5 and 10 μA)

However, B_{c2} decreases sharply for values of current density up to 2.5×10^{-3} MA/cm^2 in all isotherms due to the vortex motion and thus the energy dissipation is translated into finite resistance values. Similarly, B_{c2} decreases as a function of temperature because when we approach temperatures close to T_c, the coherence length of the superconducting material increases as described by the Ginzburg–Landau equations

$$\xi \propto \frac{1}{(T_c - T)^{1/2}} \tag{5.2}$$

$$H_{c2} = \frac{\Phi_0}{2\pi\mu_0\xi^2} \propto 1 - \frac{T}{T_c} \tag{5.3}$$

(where ξ is the coherence length and Φ_0 the magnetic flux quantum) and the vortex density necessary to penetrate into the material to reach the normal state is smaller than at low temperature.

5.5.3 Superconducting W-Based Ultranarrow Nanowires

In this section we address the growth and characterization of superconducting W-based ultranarrow NWs grown by FIBID, with lateral size of ~50 nm. The in situ SEM images were taken using the dual beams where the samples were fabricated (Nova 200 NanoLab, Helios NanoLab 600), thus providing high-resolution images of the NWs and giving a good determination of the lateral size. Additionally, AFM measurements were made to investigate the thickness uniformity along the NWs' length.

In Fig. 5.19a we show a SEM image of the geometry used for electrical four-probe measurements. Samples M1–M6 were grown using the growth parameters $V_i = 30$ kV, $I_i = 1$ pA, time = 1 min. First, a horizontal NW of 8 μm length and four perpendicular equidistant NWs of 2 μm length used as I^+, V^+, V^-, and I^- contacts were fabricated. The distance between the voltage contacts is 2.6 μm. Then, four Pt FIBID microwires were grown using trimethylcyclopentadienylmethyl-platinum, $(CH_3)_3Pt(CpCH_3)$, as a precursor gas, connecting the W nanostructure to the Ti pads [31, 32]. The coincidence point between the FEB and FIB is perfectly aligned, so we adjusted the deposition area using FEB, thus avoiding any FIB irradiation of the deposition area. In Fig. 5.19b a SEM image of M1 is shown, with $w \sim 50$ nm, and in (c) an AFM image of M1 is shown, with $t \sim 30$ nm.

We plot the normalized resistance of M1 as a function of temperature using 0.1 μA as a bias current in Fig. 5.20a and the normalized resistance as a function of the bias current at different temperatures in Fig. 5.20b. T_c is estimated as the midpoint of the transition ($R_{normalized} = 0.5$) and the T_c of the six samples ranges from 4.2 to 4.8 K. T_c is very close to the values obtained for W nanostructures both by other researchers and by us in previous experiments, where $T_c \sim 5$ K [9, 10, 12, 16, 28–30].

In Table 5.3 we summarize the most relevant parameters of superconducting W-based ultranarrow NWs. We obtain relatively high (in the range of $\sim 10^5$ A/cm^2) critical current density J_c, as previously obtained for this material [3].

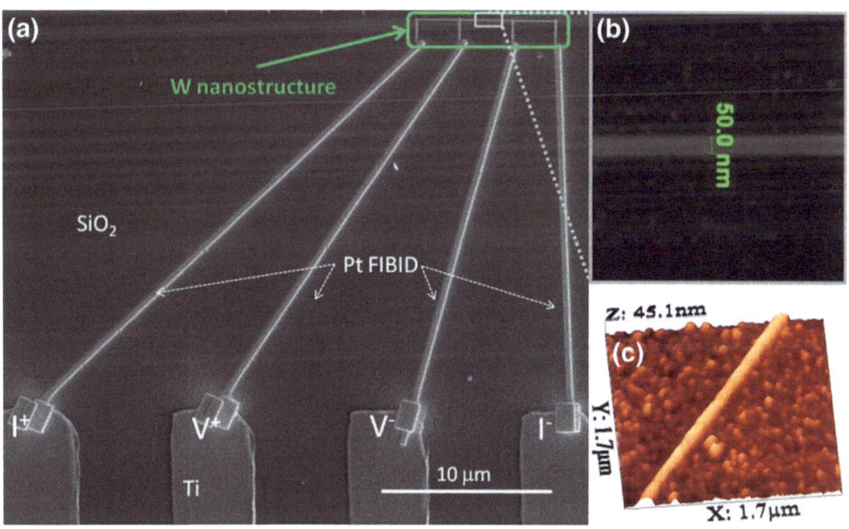

Fig. 5.19 **a** SEM image of the geometry used for four-probe electrical measurements, where the *green rectangle* indicates the W nanostructure, and the four *white arrows* indicate Pt FIBID microwires connecting to the Ti pads. **b** SEM image of M1. **c** AFM image of M1. Adapted with permission from [36]. Copyright from (2013), NJP

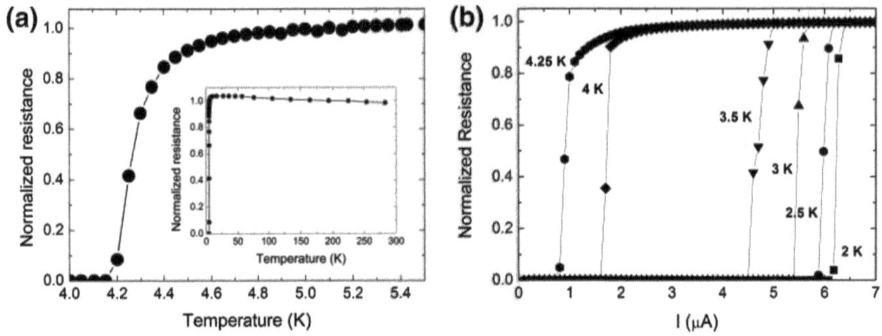

Fig. 5.20 **a** Normalized resistance as a function of temperature for M1. **b** Normalized resistance as a function of bias current for M1

Table 5.3 Dimensions of the NWs (width and thickness) and physical properties studied (resistivity at 300 K, critical transition temperature, and the critical current density at 2 K)

Sample	Width (nm)	Thickness (nm)	$\rho_{300\ K}$ (μΩ cm)	T_c (K)	$J_{c\ 2K}$ (MA/cm^2)
M1	50 ± 5	30 ± 5	172	4.25	0.41
M2	60 ± 5	49 ± 5	254	4.50	0.22
M3	60 ± 5	37 ± 5	215	4.75	0.19
M4	65 ± 5	44 ± 5	223	4.42	0.16
M5	70 ± 5	44 ± 5	276	4.37	0.15
M6	50 ± 5	30 ± 5	205	4.80	0.36

Now, we focus on the dependence of resistance on applied magnetic field in sample M1, but the same effect was shown in other nanostructures studied in this section. The magnetic field dependence *R*(*B*) of the resistance of the W-based nanowires at different temperatures is shown in Fig. 5.21 (top panel).

At $T = 4$ K, the *R*(*B*) dependence is monotonic. At lower temperatures, R(B) develops a peculiar N-shape that shifts to slightly higher fields as the temperature decreases. At the lowest temperatures, *R*(*B*) has a maximum and then drops into an extended reentrant zero-resistance region (within the noise level of 100 mΩ, high-field value is $R_N = 3.56$ kΩ). The decrease in the resistance is roughly exponential as function of temperature $R \propto \exp[T_0(B)/T]$. The corresponding activation energy $T_0(B)$ is shown in Fig. 5.21 (lower panel), and increases by a factor of three with increasing magnetic field.

To obtain insight into the mechanism of the observed non-monotonic *R*(*B*) behavior and reentrant zero-resistance state, we review the thermodynamics of a superconducting wire (strip) of width $w \geq \xi_d$. Following the calculations presented in Ref. [37] for a slab of width $w \geq \xi_d$ in a parallel field, we deduce the equilibrium phase diagram shown in Fig. 5.22a. At fields $B < B_V$, the superconducting and normal states are separated by the thermodynamic critical field $B_c = \sqrt{3}\Phi_0/(\pi\xi_d w)$, where $\Phi_0 = \pi\hbar c/e$ is the flux quantum. At high fields, $B \gg B_V$, superconductivity vanishes at the upper critical field for surface

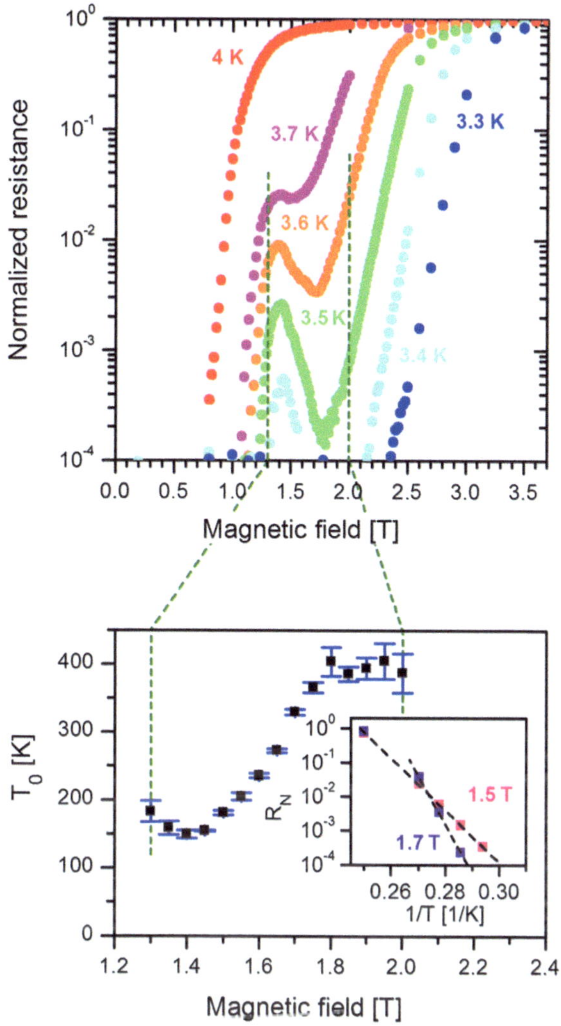

Fig. 5.21 *Upper panel* The magnetic field dependences of the resistance of the W-based nanowire measured at different temperatures. The data are normalized to high-field values for each temperature. Note the large range of fields for reentrant superconductivity. Below 3.5 K, the resistance becomes immeasurably small within nearly half a tesla. The data at $T = 3.3$ K are normalized to the high-field resistance $R_N = 3.56$ kΩ, the noise level is 100 mΩ, so the overall drop in $R(B)$ in the reentrant region is by a factor of 10^{-4}. *Lower panel* The variation of the activation energy T_0 with the magnetic field in the region of the formation of the reentrant dissipation-free state. *Inset* Representative Arrhenius plot. Reprinted with permission from [36]. Copyright from (2013), NPG

superconductivity $B_{c3} = 1.695 B_{c2}$, where $B_{c2} = \Phi_0/2\pi\xi_d^2(T)$. The field B_V divides bulk and surface superconductivity regions: At $B < B_V = 1.61\ s(\Phi_0/w^2)$, the amplitude of the superconducting order parameter is maximal at the middle line of the wire and is suppressed towards the edges, while at $B > B_V$, the situation is reversed and superconductivity exists in a surface (edge) sheath [37]. Equilibrium vortices settle in a strip at $B = B_V$ as a one-dimensional vortex row along the center [37]. Within the parameter range of our experiment, the wire accommodates only one vortex row. The point $\{T_V, B_V\}$, where $B_V = B_c(T_V)$, is a tri-critical point. We find in our wire, $B_V = 1.3$ T and $T_V = 3.87$ K. These values coincide with very good precision with the experimentally found magnetic field of the maximum in

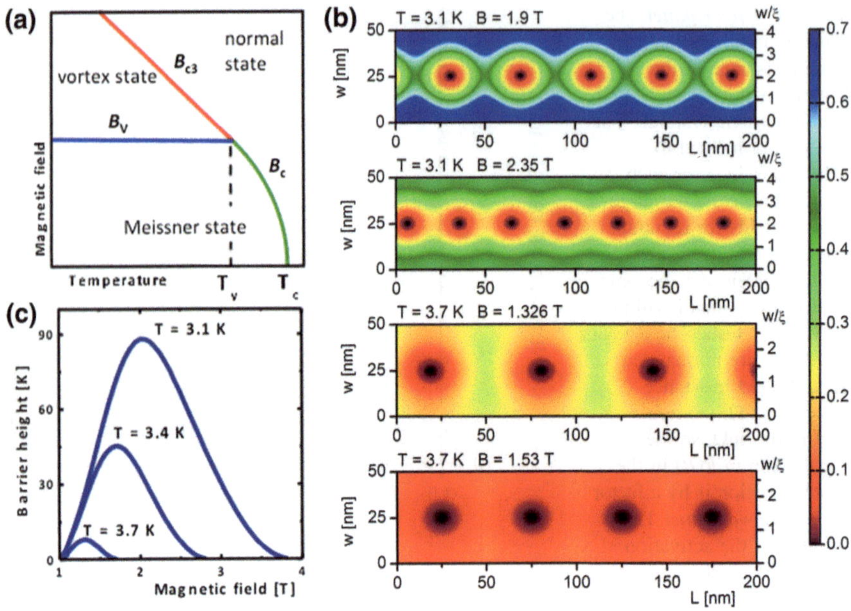

Fig. 5.22 **a** Sketch of the phase diagram obtained within the framework of the Ginzburg–Landau description of the superconducting wire with $w \geq \xi_d$. The thermodynamic magnetic field B_c separates the vortex-free state from the normal metal at $B < B_V$; at $B > B_V$ the upper surface critical field B_{c3} has this role. When $B_c = B_V$, we find a tri-critical point in the phase diagram, whose position coincides with the temperature T_V, above which vortices do not enter the wire, because the coherence length ξ_d has grown over the lateral size of the wire. Calculations give $B_V = 1.4$ T and $T_V = 3.9$ K for the wire parameters. **b** Evolution of the order parameter in the superconducting wire with the applied magnetic field. The color map of the square modulus of the order parameter $|\Psi|^2$ for representative temperatures and magnetic fields from numerical solution of the Ginzburg–Landau equations. The color scale is the same for all four panels. The *upper two panels* show the increase in linear vortex density above B_V. Temperature is too low for vortex motion in the resistance. The *two lower panels* show, respectively, the order parameter distribution at the magnetic field where barrier height is maximal, that is, close to the minimum of resistance, and close to B_{c3}. **c** The activation barrier for single vortex motion as a function of the magnetic field for $T = 3.4$ and 3.7 K. Non-monotonic behavior of the barriers reflects the competition between reduced vortex motion, and destruction of superconductivity. Adapted with permission from [36]. Copyright from (2013), NPG

$R(T)$ ($B_{max} \approx 1.4$ T) and the experimental temperature above which no maximum is observed (≈ 3.9 K). Numerical solutions of the Ginzburg–Landau equations [38] illustrate suppression of the superconducting order parameter near the middle line and onset of surface-like superconductivity at $B > B_V$ (Fig. 5.22b). The temperature above which no vortices appear in the wire (3.8 K) is also in a good agreement with T_V, which is also close to temperature defining the condition [39] $w > 1.85\xi_d(T)$ for vortices to first enter the wire.

With the phase diagram at hand, we are now equipped to explain the magnetic field dependence of the resistance $R(B)$. At $B < B_V$, the dissipation is caused by the ‘external’ vortices crossing the wire. Every vortex sees a dome-like potential barrier with the maximum at the strip center [40] and the height linearly decreasing with B. Correspondingly, $R \propto \exp(B/B^*)$ with $B^* = 0.028$ T at $T = 4$ K, in a fair agreement with the experimental data.

At $B = B_V$, a single row of equilibrium vortices settles in the middle of the wire. Thus, at $B > B_V$ the dissipation is governed by the escape of these vortices from the strip across the barriers created by edge superconductivity. At $B \approx B_V$, the newly established edge barriers are nearly zero. As they also vanish at $B = B_{c3}$, where superconductivity is fully suppressed, these barriers must achieve their maximal value at some field between B_V and B_{c3}, giving rise to the N-shape of $R(B)$ with the maximum at B_V. To quantify suppression of $R(B)$, we note that as the edge superconducting sheath at $B > B_V$ has a width of $\sim \xi_d$, a vortex crossing the edge barrier can be viewed as a phase slip (PS) in the sheath [41, 42]. To derive the PS activation energy E_{PS}, we employ the concept of the effective resistive medium. There the central band containing vortices can be considered a normal metal with the effective density of normal excitations $\nu(B) = (B/B_{c3})\nu_N$, namely the density of excitations within a single core (ν_N) times the dimensionless linear vortex density (B/B_{c3}). Then the activation energy governing the resistance is the PS nucleation energy, $E_{PS} \approx E_C F(B)$, where E_C ($= 2\nu_N \Delta^2 \xi_d^2 s$) is the condensation energy at zero field and $F(B)$ is a dimensionless function of the magnetic field such that $F(B) \sim B/B_{c3}$ at $B \ll B_{c3}$, and $F(B) \approx (1 - B/B_{c3})^{5/4}$ as $B \to B_{c3}$. In the expression for E_C, Δ is the superconducting gap and $\xi_d^2 s$ the volume of the PS nucleus ($s = \min\{\xi_d(T), d\}$, where d is the thickness of the sample). As $F(B)$ varies from ≈ 0.15 to 0.4 and $E_C(T) \approx 25 k_B T_c$ for the strip, the resistance $R(B) \propto \exp(-E_{PS}/T)$ is indeed effectively suppressed over a large field interval between B_V and B_{c3} at $T < 3.5$ K. The remarkable effect of the increase of E_{PS} with the magnetic field reflects the fact that, at $B \geq B_V$, the ‘metallicity’ of the resistive central layer of the wire (that is, the effective density of the normal excitations near the center) increases faster than the suppression of surface superconductivity at the edges. The field-induced metallicity means merely a slight suppression of the order parameter along the center by an increasingly dense vortex chain. Therefore, densely packed vortices dig a deeper and deeper potential well for themselves with increasing magnetic field and stop their transverse motion. The current is short circuited by the edge superconducting channels and the wire turns superconducting. Upon further increase of magnetic field, the suppression of edge superconductivity takes over and $R(B)$ starts to grow again.

That the escape barriers for a vortex chain in a narrow superconducting slab can grow with the magnetic field and lead to enhancement of superconducting parameters was discussed in [43]. Unfortunately, while correctly capturing the trend, the image approach used in [43] did not take into account the collective suppression of the order parameter near the center and thus underestimated the magnitude of the escape barrier. As a result, the effect came out significantly reduced. More precise Ginzburg–Landau simulations accounting for the self-induced suppression of the order parameter along the vortex chain show a gigantic

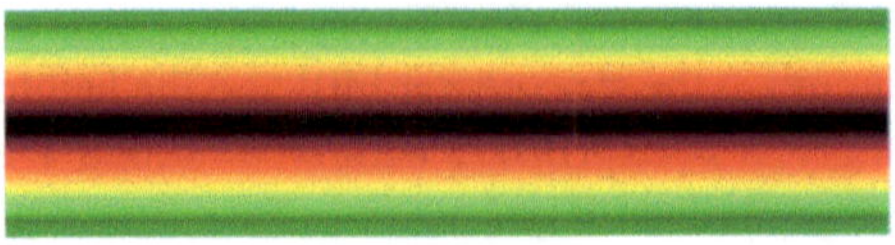

Fig. 5.23 Sketch of the nearly tri-layered state leading to recovered superconductivity in a wire. The inner (*black*) stripe is a nearly metallic state and the green layers show the edge-enhanced superconductivity. Adapted with permission from [36]. Copyright from (2013), NPG

increase of the escape barrier with the magnetic field, agreeing with the effective-medium-based estimates (Figs. 5.22b, c and 5.23).

We have demonstrated giant magnetic field–induced reentrant superconductivity in a thin W-based wire. The giant reentrance rests on two components. The first is that surface (edge) superconductivity stops vortex motion by constricting them into immobile aggregates: the denser the vortex cluster, the better the superconducting order parameter around this cluster is suppressed and the deeper the self-generated potential well confining the vortices. The effective resistive medium grows more metallic with increasing field. The density of normal excitation states is proportional to B, and so is the energy of the PS nucleus. Thus, increasing magnetic field suppresses the phase slips across the edge superconducting channels until the suppression of surface superconductivity itself prevails. The second component is the specific geometry ensuring that the volume occupied by edge superconducting channels is comparable to that of the vortex-carrying resistive medium. In that case, collective depinning effects do not develop, implying that the surface barrier for vortex escape can stand applied currents comparable to the pair-breaking current. Our findings open new routes for the design of superconducting devices working at high magnetic fields and temperatures, and mark an important step towards resolving the long-standing problem of extending the coherent dissipation-free superconducting state to the high field and temperature range.

A similar behavior of the N-shaped curves decreasing bias current at 0.25 and 0.1 μA and even increasing to 1 μA have been observed, in which the minimum resistance was observed in the range of 3.2–3.7 K. In order to verify that the reentrance of the dissipation-free state does not depend on the measurement mode, or the polarity of the applied magnetic field, measurements were performed in AC and DC modes from +9 to −9 T over a range of temperatures.

A study of the vortex dynamics in these nanostructures by STM/S would provide a direct image of the vortex arrangement and shed more light for the purposes of the reentrance of the dissipation-free state. However, this interesting study is extremely complex and has not been successfully addressed to date.

5.5.4 Summary and Conclusions

In Table 5.4 we summarize the deposition parameters and the superconducting properties of W-based structures studied by electrical measurements in local mode.

Table 5.4 Summary of deposition parameters, dimensions (width and thickness), resistivity at 300 K, and superconducting properties (critical transition temperature, critical current density, and upper critical field) of W-based superconducting structures studied by electrical measurements in local mode

	Microwires	Nanowires	Ultranarrow nanowires
V_i (kV)	30	30	30
I_i (nA)	0.92	0.029	0.001
Width (nm)	2000	200	50
Thickness (nm)	300	200	30
$\rho_{300\ K}$ (μΩ cm)	214	255	172
T_c (K)	4.95	5.10	4.25
J_c (MA/cm^2)	0.013 (2.5 K)	0.094 (1 K)	0.41 (2 K)
$B_{c2,\ 2\ K}$ (T)	~7	8.44	4.72

In addition to the reentrance of the dissipation-free state observed in superconducting ultranarrow NWs and explained in detail in the previous section, the fundamental differences detected from structures of different lateral width are J_c and $B_{c2,2K}$. The critical current density without external magnetic field in superconducting W ultranarrow NWs is one order of magnitude higher than the values obtained in both microwires and nanowires. In a bulk material, as W nanowires and microwires may be considered, the critical current density is governed by the pinning forces. However, if the dimensions of the structures are reduced to values close to ξ, superconductivity is preserved at the edges of the structures, so high current density is needed to be applied to overcome the energy barrier [44–46]. However, the upper critical field in the superconducting W ultranarrow NWs is at least a factor of 1.5 lower than the results obtained in W nanowires and microwires. This behavior would be given by the energy dissipation mechanism, which holds exclusively in this kind of ultranarrow nanostructures in which width or thickness are of the same order as the coherence length. In this case, the phase slip phenomenon is the mechanism that governs the energy dissipation up to 3.8 K.

5.6 Nonlocal Electric Transport in Superconducting W-Based Deposits

In this section, we address the nonlocal electrical transport in superconducting W NWs grown by FIBID, with width and thickness of ~200 nm as shown in Sect. 5.2.

Thus, in type II superconductors with weak pinning subjected to an applied magnetic field perpendicular to the plane, an applied current in the upper horizontal NW that acts on the vortices located in the upper cross, pushing in the direction of the vertical NW due to the Lorentz force, generates a nonlocal voltage in the lower cross (see Figs. 5.24 and 1.20). The repulsion between vortices,

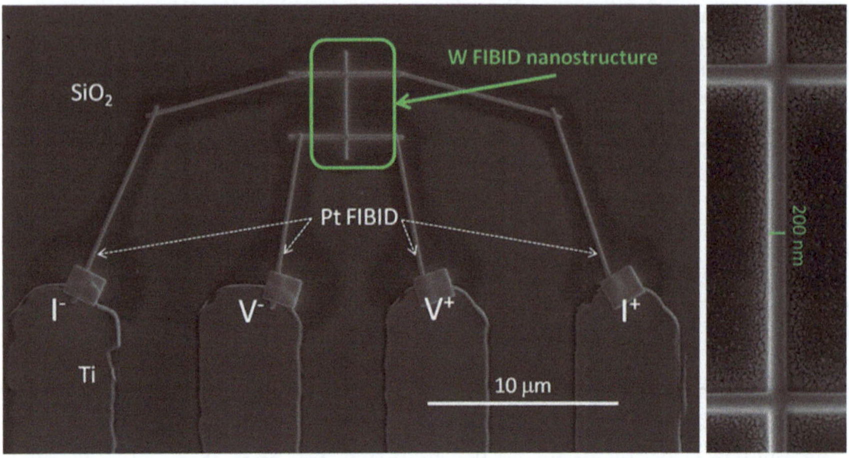

Fig. 5.24 *Left panel* SEM image of the geometry used in nonlocal electrical measurements. The additional four Pt FIBID lines were contacted to Ti pads after the W-based nanostructure (marked by the *green rectangle*) was deposited to enable four-point measurements to be made. *Right panel* SEM image of the W-based nanostructure

combined with the vortex density determined by the external magnetic field, leads to an efficient pressure transfer generated in the vortex lattice through the vertical NW, and once vortices are moved towards the lower cross, a nonlocal voltage is generated by the relation with the electric field (Eq. 1.7), $\vec{E} = \vec{B} \times \vec{v}_{\phi}$.

The model proposed in references [47–50] explains very well the observed behavior of the linear nonlocal signal. Taking into account the sample geometry (with length L, width w, and thickness t), and the vortex density $n_{\Phi} = B/\Phi_0$, there are a number of vortices $N = n_{\Phi} w^2$ on the top cross, which feel the Lorentz force $F_L = J\Phi_0 t$ as a result of the applied current density $J = I/wt$, thus leading to the following pressure on the vortices:

$$p = \frac{n_{\Phi}\Phi_0 I}{t} \tag{5.4}$$

The resultant force on vortices per unit length is $F = pw$; equating this with the frictional force ($F_f = n_{\Phi} L w \eta v_{\varphi}$) to move vortices and using $V_{nonlocal} = wBv_{\varphi}$, where η and v_{φ} are the viscosity and the vortex velocity in the NW, respectively, the resulting expressions are

$$V_{nonlocal} = \frac{wB\Phi_0 I}{\eta L t} \tag{5.5}$$

This model explains very well the observed behavior of the nonlocal signal. The pinning force prevents the vortex motion below a magnetic field threshold, in both the local and the nonlocal signals. Our amorphous and weak-pinning material seems to have a strong density of pinning centers, which are first used up for vortex nucleation then, when they are saturated, vortices can move along the easy axis. In fact the nonlocal signal can propagate long distances—up to micrometers—compared to the vortex distance. This effect can be attributed to the rigidity of the vortex lattice. In this mechanism, the vortices' confinement along the NW could be due to the surface pinning at its edges. Measurements in narrow NWs with less effective pressure transfer point to a region of about 50 nm where vortices are strongly pinning at the edges, so flux paths in the sample are formed [51].

5.6.1 Electrical Measurement Geometry

A SEM image of the nonlocal geometry of W-based nanostructures grown by FIBID is shown in Fig. 5.24. The deposition parameters are $V_i = 30$ kV, $I_i = 29$ pA, and the dimensions are 200 nm in width and thickness, and 3 μm in length between the bias current contacts and the voltage contacts. Then, four Pt–C microwires were deposited using FIBID with $(CH_3)_3Pt(CpCH_3)$ precursor gas, to connect the W nanostructure with the Ti lines, using suitable deposition parameters to guarantee a good contact [31, 32].

5.6.2 Bias Current Dependence of the Nonlocal Voltage

Systematically, nonlocal voltage measurements were made as a function of the bias current (from 0 to 40 μA) in AC mode at low temperatures (0.5, 1, 2 K) and as a function of applied magnetic field (from 1 to 9 T). The dependence of the nonlocal voltage on the bias current at 0.5 K and at applied magnetic field as indicated in the legend is shown in Fig. 5.25. As an example of the generated nonlocal signal, we focus on the curve at 4 T, as clearly there are three different

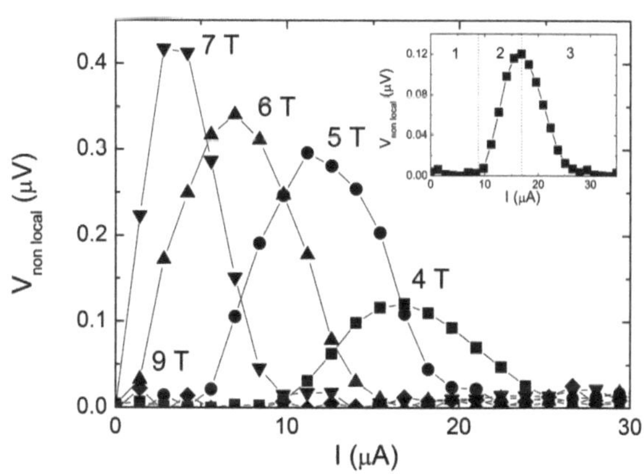

Fig. 5.25 Bias current dependence of the nonlocal voltage at 0.5 K and at different external magnetic fields as indicated. *Inset* Bias current dependence of the nonlocal voltage at 0.5 K and 4 T; the three regimes of the nonlocal voltage explained in the main text are indicated

regions (see inset of Fig. 5.25). In region 1, the nonlocal signal does not exist from 0 to 10 μA; this effect could be related to the pinning forces. In region 2, we observe a linear dependence of $V_{nonlocal}$ on the applied current, as expected for a Lorentz force. Finally, $V_{nonlocal}$ decreases monotonically in region 3, from 17 μA onwards, until it cancels, which indicates that the critical current value is exceeded. In the next paragraphs, we will discuss the effects involved in each region.

The linear nonlocal signal observed in Fig. 5.25 and its inset corresponding to the region 2 would be consistent with the model previously described by Eq. (5.4). The efficient transfer of pressure occurs in the vertical NW, which moves the rigid vortex lattice until it reaches the lower cross, where the nonlocal voltage is measured. The origin of the pressure is the Lorentz force that builds up in the upper cross where electric current is applied (see Fig. 1.19). However, we must have a minimum vortex density in the NW to allow efficient transfer of pressure and also to overcome the pinning in the vortex lattice to move it along the vertical NW. For instance, at 0.5 K the nonlocal signal is observed at magnetic field values up to 4 T, as shown in Fig. 5.25. Thus, the maximum nonlocal signal increases as a function of the applied magnetic field, until it cancels at 9 T, when the NW reaches the normal state. We should note that the nonlocal signal is not observed in all cases when the current is applied in the upper horizontal NW, contrary to cases in the literature [47, 49, 50]. Each applied field value gives us a vortex density, which first nucleates and then, when the pinning centers are saturated, the vortex lattice has a different pinning force that is exceeded by using the applied current as the tool to initiate the propagation mechanism of the vortex motion along the vertical channel. The maximum nonlocal voltage was observed at high applied magnetic field without exceeding B_{c2}.

In addition, the decrease of $V_{nonlocal}$ in each curve at a given current value is associated with values close to the critical current at a particular temperature and applied magnetic field.

On the other hand, the effect of the polarity of the current on $V_{nonlocal}$ was studied by means of DC mode measurements. In Fig. 5.26 we present $V_{nonlocal}$ as a

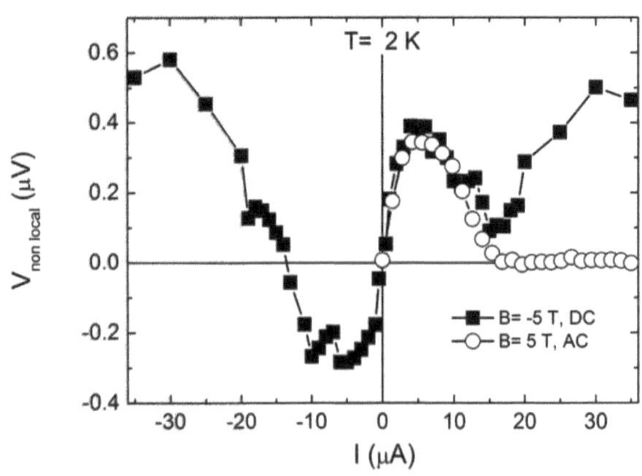

Fig. 5.26 Bias current dependence of the nonlocal voltage at 2 K and −5 T magnetic field using the DC measurement mode, and at 5 T using the AC measurement mode

function of the current from 35 μA to −35 μA at 2 K and −5 T measured in DC mode and also $V_{nonlocal}$ from 0 to 35 μA at 2 K and 5 T measured in AC mode. First, we highlight the asymmetry of the nonlocal signal and its linear dependence under current from 5 μA to −5 μA, which follows the same dependence as the measurement made in the AC mode. If we go on increasing the applied current, the voltage becomes a nonlocal symmetric signal that does not depend on the polarity of the current and therefore the results in AC mode cancel. As current is applied only in the upper horizontal NW, we expect to generate a gradient of temperature along the vertical channel between that NW and the lower horizontal NW, which could imply the existence of the nonlocal signal due to a temperature gradient such as in DC mode measurement, once the upper cross was in the normal state.

Thus, we have demonstrated that the nonlocal signal obtained from the AC mode measurements does not introduce a mistake signal and does not hide any relevant data in our nonlocal signal.

5.6.3 Applied Magnetic Field Dependence of the Nonlocal Resistance

We studied $R_{nonlocal}$ as a function of magnetic field (± 9 T) perpendicular to the substrate and temperatures from 0.5 to 5 K in W NWs.

The nonlocal signal generated at constant temperature and current appears at a magnetic field value, which can be associated with the depinning of the vortex lattice, which is free to move; and specifically with the saturation of pinning centers of the interstitial vortices at high vortex densities. Then the maximum nonlocal resistance is reached around a fixed field value, and it is reduced to zero when B_{c2} is approached. As an example, in Fig. 5.27a we show $R_{nonlocal}$ and normalized R_{local} as a function of the magnetic field at 2 K and with 0.5 μA applied current. In addition, $R_{nonlocal}$ in DC mode as a function of the magnetic field at 2 K for 1, 5, and 10 μA applied currents is shown in Fig. 5.27b.

In Fig. 5.28, we plot $R_{nonlocal}$ as a function of the magnetic field from 0.5 to 5 K with 1 μA applied current. A successive offset in $R_{nonlocal}$ has been applied, to improve the clarity of the graph. The dependence of $R_{nonlocal}$ is symmetric at positive and negative values of the magnetic field. The maximum $R_{nonlocal}$ (0.32 Ω at 0.5 K) extracted from Fig. 5.28 decreases as a function of the temperature from 0.5 K to 3 K ($t = T/T_c = [0.1–0.6]$) as we can see in Fig. 5.29a. This could be explained as a consequence of the B_{max} growing faster with temperature than the frictional coefficient does [47]. However, the maximum nonlocal signal does not change significantly up to 3 K and cancels at 5 K, $t = 0.98$. This effect could be explained as due to the confinement of vortices by surface barriers, once the vortex lattice has melted at temperatures close to T_c. From Figs. 5.28 and 5.29b, the nonlocal signal appears at $b_{min} = B_{min}(T)/B_{c2}(T) = [0.35–0.64]$, attains its maximum value at $b_{max} = B_{max}(T)/B_{c2}(T) = [0.46–0.82]$, and after this value decreases until it cancels.

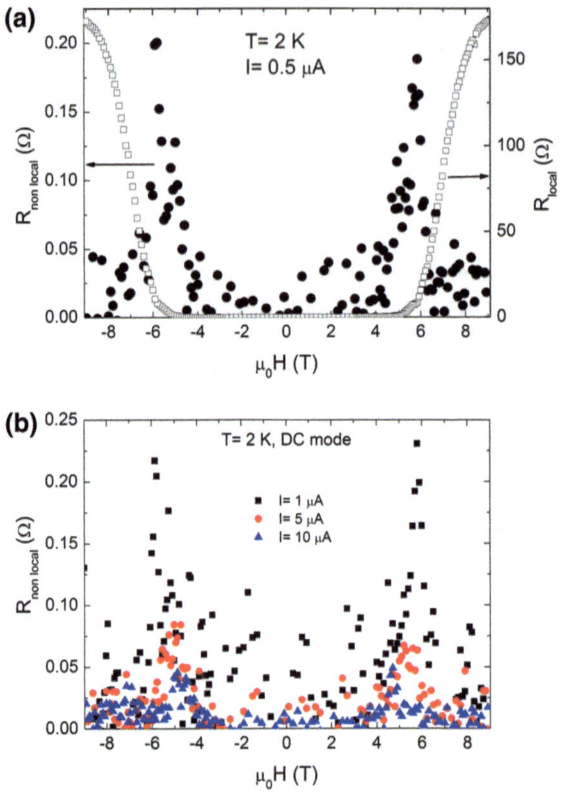

Fig. 5.27 **a** Magnetic field dependence of the nonlocal resistance (*left* *y*-axis) together with the corresponding normalized local resistance from the local measurement (*right* *y*-axis) at 2 K and 0.5 μA bias current. **b** Magnetic field dependence of the nonlocal resistance in DC mode at 2 K and 1, 5, and 10 μA bias currents

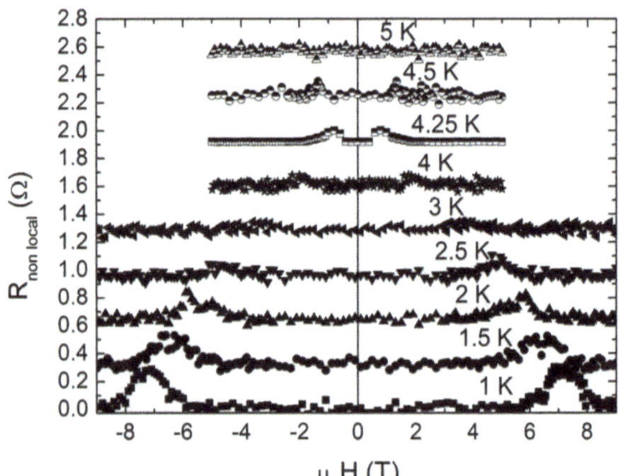

Fig. 5.28 Magnetic field dependence of the nonlocal resistance at 1 μA bias current and for different temperatures as indicated. An offset in the R_{nonlocal} has been applied successively to give a clearer graph

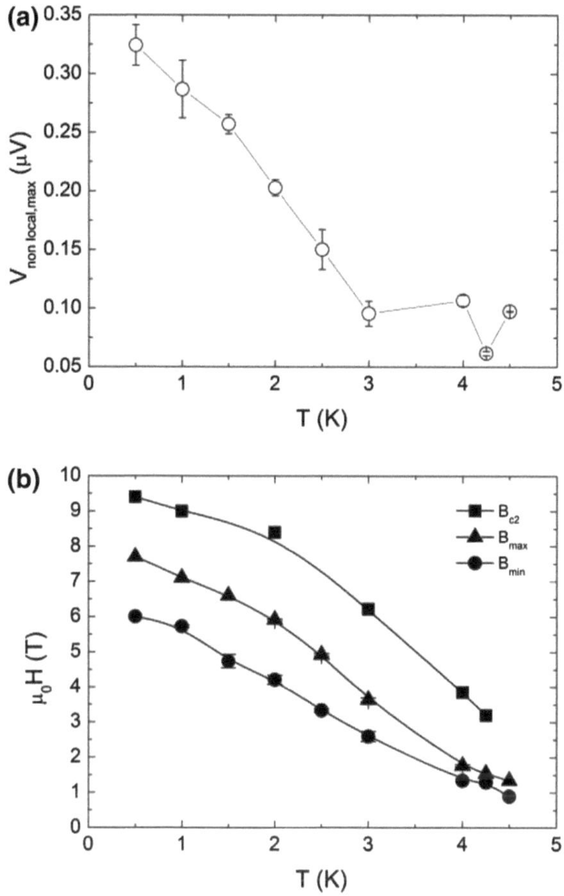

Fig. 5.29 **a** Temperature dependence of the maximum nonlocal voltage [$V_{\text{nonlocal,max}} = (V_{\text{nonlocal,max}}(+B) + V_{\text{nonlocal,max}}(-\text{B}))/2$]. **b** Temperature dependence of the magnetic fields indicated in the legend [$B_{\max} = (B_{\max}(+B) + B_{\max}(-\text{B}))/2$; $B_{\min} = (B_{\min}(+B) + B_{\min}(-\text{B}))/2$]

We carried out a thorough investigation of the nonlocal signal detected in superconducting W FIBID NWs by means of magnetotransport measurements. The nonlocal signal is caused by the motion of the vortex lattice in our low-pinning superconducting NW, which is based on a pressure transfer in the NW, where current is not applied.

Besides, the nonlocal signal was observed in all ranges of the superconductivity studied (temperature, $t = [0.10\text{–}0.83]$; external magnetic field, $b = [0.35\text{–}0.82]$; and current density, $J(\text{MA/m}^2) \sim [2.5\text{–}750]$), while exceeding the pinning and frictional forces. We highlight a dependence of the nonlocal signal on the temperature, which is explained by the interaction between the vortex density and the frictional force of the vortex lattice.

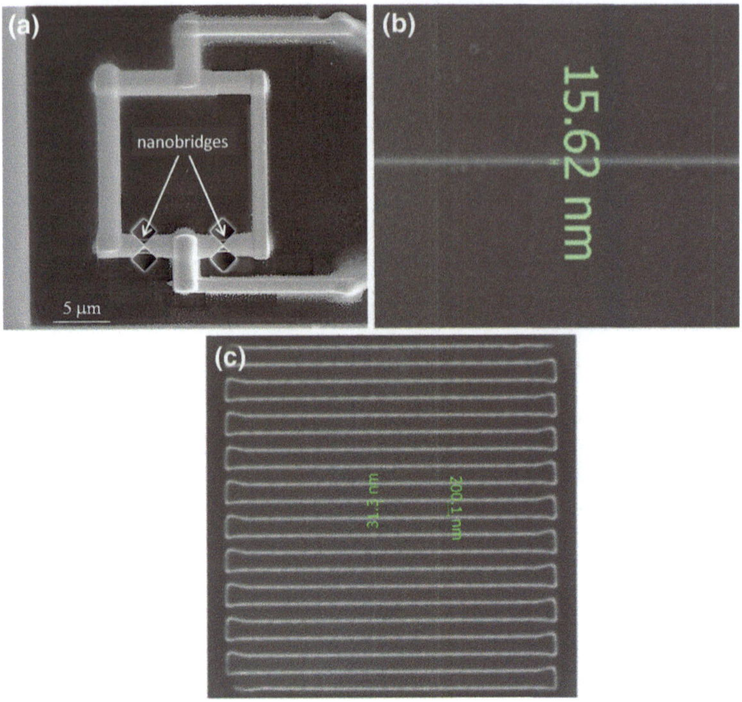

Fig. 5.30 SEM images of W superconducting structures grown by FIBID: **a** a microSQUID, **b** an ultranarrow NW, and **c** a serpentine for a SSPD

5.7 Conclusions and Outlook

In this chapter we report how W deposits grown by FIBID using $W(CO)_6$ as a precursor material were studied by several techniques, such as EDS, HRTEM, and magnetotransport measurements in both local and nonlocal modes.

The control of growth and design of superconducting materials at the nanoscale is opening up interesting application fields in condensed matter physics. Recently, studies in superconductivity in one dimension [52], vortex confinement [53, 54], vortex pinning [55], etc. have been tackled. Fabrication of nanoSQUIDs [56, 57], three-dimensional structures, as well as support material for superconducting computers (superconducting qubits, SQBs) [58], fabrication of superconducting single-photon detectors (SSPDs) [59, 60] or particles [61] and quantum junctions based on quantum phase slip centers (QPSCs) [62] are some of the attractive applications of such nano-superconductors. In Fig. 5.30 some examples of the prospects offered by these nanostructures are shown.

References

1. Langfischer, H., Basnar, B., Hutter, H., Bertagnolli, E.: Evolution of tungsten film deposition induced by focused ion beam. J. Vac. Sci. Technol. A—Vac. Surf. Films, **20**(4), 1408–1415 (2002)
2. Muthukumar, K., Opahle, I., Shen, J., Jeschke, H. O., Valenti, R.: Interaction of $W(CO)_6$ with SiO_2 surfaces: a density functional study. Phys. Rev. B, **84**(20), 205442 (2011)
3. Sadki, E.S., Ooi, S., Hirata, K.: Focused-ion-beam-induced deposition of superconducting nanowires. Appl. Phys. Lett. **85**(25), 6206–6208 (2004)
4. Collver, M.M., Hammond, R.H.: Superconductivity in amorphous transition-metal alloy films. Phys. Rev. Lett. **30**(3), 92–95 (1973)
5. Kondo, S.: Superconducting characteristics and the thermal-stability of tungsten-based amorphous thin-films. J. Mater. Res. **7**(4), 853–860 (1992)
6. Miki, H., Takeno, T., Takagi, T., Bozhko, A., Shupegin, M., Onodera, H., Komiyama, T., Aoyama, T.: Superconductivity in W-containing diamond-like nanocomposite films. Diam. Relat. Mater. **15**(11–12), 1898–1901 (2006)
7. Osofsky, M.S., Soulen, R.J., Claassen, J.H., Trotter, G., Kim, H., Horwitz, J.S.: New insight into enhanced superconductivity in metals near the metal–insulator transition. Phys. Rev. Lett., **87**(19), 197004 (2001)
8. Tinkham, M.: Introduction to Superconductivity, 2nd edn. Dover Publications, Inc., New York (1996)
9. Luxmoore, I.J., Ross, I.M., Cullis, A.G., Fry, P.W., Orr, J., Buckle, P.D., Jefferson, J.H.: Low temperature electrical characterisation of tungsten nano-wires fabricated by electron and ion beam induced chemical vapour deposition. Thin Solid Films **515**(17), 6791–6797 (2007)
10. Spoddig, D., Schindler, K., Roediger, P., Barzola-Quiquia, J., Fritsch, K., Mulders, H., Esquinazi, P.: Transport properties and growth parameters of PdC and WC nanowires prepared in a dual-beam microscope. Nanotechnology, **18**(49), 495202 (2007)
11. Li, W.X., Fenton, J.C., Wang, Y.Q., McComb, D.W., Warburton, P.A.: Tunability of the superconductivity of tungsten films grown by focused-ion-beam direct writing. J. Appl. Phys., **104**(9), 093913 (2008)
12. Li, W., Fenton, J.C., Warburton, P.A.: Focused-ion-beam direct-writing of ultra-thin superconducting tungsten composite films. IEEE Trans. Appl. Supercond. **19**(3), 2819–2822 (2009)
13. Martínez-Pérez, M. J., Sesé, J., Córdoba, R., Luis, F., Drung, D., Schuring, T.: Circuit edit of superconducting microcircuits. Supercond. Sci. Technol. **22**(12), 125020 (2009)
14. Martínez-Pérez, M.J., Sesé, J., Luis, F., Córdoba, R., Drung, D., Schurig, T., Bellido, E., de Miguel, R., Gomez-Moreno, C., Lostao, A., Ruiz-Molina, D.: Ultrasensitive broad band SQUID microsusceptometer for magnetic measurements at very low temperatures. IEEE Trans. Appl. Supercond. **21**(3), 345–348 (2011)
15. Martínez-Pérez, M. J., Sesé, J., Luis, F., Drung, D., Schurig, T.: Highly sensitive superconducting quantum interference device microsusceptometers operating at high frequencies and very low temperatures inside the mixing chamber of a dilution refrigerator. Rev. Sci. Instr. **81**(1), 016108 (2010)
16. Li, W., Fenton, J.C., Gu, C., Warburton, P.A.: Superconductivity of ultra-fine tungsten nanowires grown by focused-ion-beam direct-writing. Microelectron. Eng. **88**(8), 2636–2638 (2011)
17. Cui, A., Li, W., Luo, Q., Liu, Z., Gu, C.: Freestanding nanostructures for three-dimensional superconducting nanodevices. Appl. Phys. Lett. **100**(14), 143106 (2012)
18. Guillamón, I., Suderow, H., Vieira, S., Fernández-Pacheco, A., Sesé, J., Córdoba, R., De Teresa, J.M., Ibarra, M.R.: Nanoscale superconducting properties of amorphous W-based deposits grown with a focused-ion-beam. New J. Phys. **10**(9), 093005 (2008). doi:10.1088/1367-2630/10/9/093005

19. Guillamón, I., et al.: Superconducting density of states at the border of an amorphous thin film grown by focused-ion-beam. J. Phys.: Conf. Ser. **150**(5), 052064 (2009)
20. Guillamón, I., Suderow, H., Fernández-Pacheco, A., Sesé, J., Córdoba, R., De Teresa, J.M., Ibarra, M.R., Vieira, S.: Direct observation of melting in a two-dimensional superconducting vortex lattice. Nat. Phys. **5**(9), 651–655 (2009). doi:10.1038/NPHYS1368
21. Guillamón, I., Suderow, H., Vieira, S., Sesé, J., Córdoba, R., De Teresa, J.M., Ibarra, M.R.: Direct observation of stress accumulation and relaxation in small bundles of superconducting vortices in tungsten thin films. Phys. Rev. Lett. **106** (7), 077001 (2011)
22. Kes, P.H., Tsuei, C.C.: Two-dimensional collective flux pinning, defects, and structural relaxation in amorphous superconducting films. Phys. Rev. B **28**(9), 5126–5139 (1983)
23. Guillamón, I.: Orden y desorden en superconductividad. Thesis, Universidad Autónoma de Madrid (2009)
24. Ziegler, J.F.: SRIM-2003. Nucl. Instrum. Methods Phys. Res. Sect. B **219–220**, 1027–1036 (2004)
25. Tripathi, S.K., Shukla, N., Kulkarni, V.N.: Exploring a new strategy for nanofabrication: deposition by scattered Ga ions using focused ion beam. Nanotechnology **20**(7), 075304 (2009)
26. Giannuzzi, L.A., Stevie, F.A.: Introduction to Focused Ion Beams, p. 357. Springer, Boston (2005)
27. Sychugov, I., Nakayama, Y., Mitsuishi, K.: Manifold enhancement of electron beam induced deposition rate at grazing incidence. Nanotechnology **21** (2), 025303 (2010)
28. Li, J.T., Toth, M., Tileli, V., Dunn, K.A., Lobo, C.J., Thiel, B.L.: Evolution of the nanostructure of deposits grown by electron beam induced deposition. Appl. Phys. Lett. **93** (2), 23130 (2008)
29. Romans, E.J., Osley, E.J., Young, L., Warburton, P.A., Li, W.: Three-dimensional nanoscale superconducting quantum interference device pickup loops. Appl. Phys. Lett. **97** (22), 222506 (2010)
30. De Teresa, J. M., Fernández-Pacheco, A., Córdoba, R., Sesé, J., Ibarra, M.R., Guillamón, I., Suderow, H., Vieira, S.: Transport properties of superconducting amorphous W-based nanowires fabricated by focused-ion-beam-induced-deposition for applications in Nanotechnology. Mater. Res. Soc. Symp. Proc. 1180 (2009)
31. Fernández-Pacheco, A., De Teresa, J.M., Córdoba, R., Ibarra, M. R.: Metal–insulator transition in Pt-C nanowires grown by focused-ion-beam-induced deposition. Phys. Rev. B **79** (17), 174204 (2009)
32. De Teresa, J.M., Córdoba, R., Fernández-Pacheco, A., Montero, O., Strichovanec, P., Ibarra, M.R.: Origin of the difference in the resistivity of as-grown focused-ion- and focused-electron-beam-induced Pt Nanodeposits. J. Nanomater. **2009**, 936863 (2009)
33. Kunchur, M.N.: Unstable flux flow due to heated electrons in superconducting films. Phys. Rev. Lett. **89**(13), 137005 (2002)
34. Babic, D.: Amorphous Nb-Ge thin films as a model system for experiments on fundamental properties of vortex transport. In: Martins, B.S. (ed.) New Frontiers in Superconductivity Research, pp. 107–143. Nova Science Publishers, Hauppauge (2006)
35. Babic, D., Bentner, J., Surgers, C., Strunk, C.: Flux-flow instabilities in amorphous $Nb_{0.7}Ge_{0.3}$ microbridges. Phys. Rev. B **69**(9), 092510 (2004)
36. Córdoba, R., Baturina, T.I., Sesé, J., Mironov, A.Y., De Teresa, J.M., Ibarra, M.R., Nasimov, D.A., Gutakovskii, A.K., Latyshev, A.V., Guillamón, I., Suderow, H., Vieira, S., Baklanov, M.R., Palacios, J.J., Vinokur, V.M.: Magnetic field-induced dissipation-free state in superconducting nanostructures. Nat. Commun. **4**, 1437 (2013). doi:10.1038/ncomms2437
37. Saint-James, D., De Gennes, P.: Onset of superconductivity in decreasing fields. Phys. Lett. **7**, 306 (1963)
38. Palacios, J.J.: Vortex lattices in strong type-II superconducting two-dimensional strips. Phys. Rev. B **57**(17), 10873–10876 (1998)
39. Fink, H.J.: Superconducting surface sheath of a type-2 superconductor below upper critical field H_{c2}. Phys. Rev. Lett. **14**(9), 309 (1965)

40. Kuit, K.H., Kirtley, J.R., van der Veur, W., Molenaar, C.G., Roesthuis, F.J.G., Troeman, A.G.P., Clem, J.R., Hilgenkamp, H., Rogalla, H., Flokstra, J.: Vortex trapping and expulsion in thin-film $YBa_2Cu_3O_{7-\delta}$ strips. Phys. Rev. B **77**(13), 134504 (2008)
41. Aranson, I., Vinokur, V.: Surface instabilities and vortex transport in current-carrying superconductors. Phys. Rev. B **57**(5), 3073–3083 (1998)
42. Anderson, P.W.: Considerations on flow of superfluid helium. Rev. Mod. Phys. **38**(2), 298–310 (1966)
43. Takacs, S.: Properties of superfine superconducting filaments embedded in normal matrix. Czech J. Phys. **36**(4), 524–536 (1986)
44. Tahara, S., Anlage, S.M., Halbritter, J., Eom, C.-B., Fork, D.K., Geballe, T.H., Beasley, M.R.: Critical currents, pinning, and edge barriers in narrow $YBa_2Cu_3O_{7-\delta}$ thin films. Phys. Rev. B **41**(16), 11203–11208 (1990)
45. Jones, W.A., Barnes, P.N., Mullins, M.J., Baca, F.J., Emergo, R.L.S., Wu, J., Haugan, T.J., Clem, J.R.: Impact of edge-barrier pinning in superconducting thin films. Appl. Phys. Lett. **97** (26), 262503 (2010)
46. Elistratov, A.A., Vodolazov, D.Y., Maksimov, I.L., Clem, J.R.: Field-dependent critical current in type-II superconducting strips: combined effect of bulk pinning and geometrical edge barrier. Phys. Rev. B **66**(22), 220506 (2002)
47. Helzel, A., Kokanovic, I., Babic, D., Litvin, L.V., Rohlfing, F., Otto, F., Surgers, C., Strunk, C.: Nonlocal vortex motion in mesoscopic amorphous $Nb_{0.7}Ge_{0.3}$ structures. Phys. Rev. B **74** (22), 220510 (2006)
48. Otto, F.: Nonlinear vortex transport in mesoscopic channel of amorphous NbGe. Thesis, Universitätsverlag Regensburg (2009)
49. Otto, F., Bilusic, A., Babic, D., Vodolazov, D.Y., Suergers, C., Strunk, C.: Nonlocal versus local vortex dynamics in the transversal flux transformer effect. Phys. Rev. B **81** (17), 174521 (2010)
50. Otto, F., Bilusic, A., Babic, D., Vodolazov, D.Y., Suergers, C., Strunk, C.: Reversal of nonlocal vortex motion in the regime of strong nonequilibrium. Phys. Rev. Lett. **104**(2), 027005 (2010)
51. Córdoba, R., Sesé, J., Ibarra, M.R., Guillamón, I., Suderow, H., Vieira, S., De Teresa, J.M.: Non local voltage in W-based nanowires grown by Focused Ion Beam Induced Deposition, manuscript in preparation
52. Arutyunov, K.Y., Golubev, D.S., Zaikin, A.D.: Superconductivity in one dimension. Phys. Rep.—Rev. Sect. Phys. Lett. **464**(1–2), 1–70 (2008)
53. Chibotaru, L.F., Ceulemans, A., Bruyndoncx, V., Moshchalkov, V.V.: Symmetry-induced formation of antivortices in mesoscopic superconductors. Nature **408**(6814), 833–835 (2000)
54. Grigorieva, I.V., Geim, A.K., Dubonos, S.V., Novoselov, K.S., Vodolazov, D.Y., Peeters, F.M., Kes, P.H., Hesselberth, M.: Long-range nonlocal flow of vortices in narrow superconducting channels. Phys. Rev. Lett. **92**(23), 237001 (2004)
55. Velez, M., Martin, J.I., Villegas, J.E., Hoffmann, A., Gonzalez, E.M., Vicent, J.L., Schuller, I.K.: Superconducting vortex pinning with artificial magnetic nanostructures. J. Magn. Magn. Mater. **320**(21), 2547–2562 (2008)
56. Foley, C.P., Hilgenkamp, H.: Why NanoSQUIDs are important: an introduction to the focus issue. Supercond. Sci. Technol. **22**(6), 064001 (2009)
57. Hao, L., Macfarlane, J.C., Gallop, J.C., Cox, D., Beyer, J., Drung, D., Schurig, T.: Measurement and noise performance of nano-superconducting-quantum-interference devices fabricated by focused ion beam. Appl. Phys. Lett. **92**(19), 192507 (2008)
58. Clarke, J., Wilhelm, F.K.: Superconducting quantum bits. Nature **453**(7198), 1031–1042 (2008)
59. Gol'tsman, G.N., Okunev, O., Chulkova, G., Lipatov, A., Semenov, A., Smirnov, K., Voronov, B., Dzardanov, A., Williams, C., Sobolewski, R.: Picosecond superconducting single-photon optical detector. Appl. Phys. Lett. **79**(6), 705–707 (2001)
60. Najafi, F., Marsili, F., Dauler, E., Molnar, R.J., Berggren, K.K.: Timing performance of 30-nm-wide superconducting nanowire avalanche photodetectors. Appl. Phys. Lett. **100**(15), 152602 (2012)

61. Sclafani, M., Marksteiner, M., Keir, F.M., Divochiy, A., Korneev, A., Semenov, A., Gol'tsman, G., Arndt, M.: Sensitivity of a superconducting nanowire detector for single ions at low energy. Nanotechnology **23**(6), 065501 (2012)
62. Mooij, J.E., Nazarov, Y.V.: Superconducting nanowires as quantum phase-slip junctions. Nat. Phys. **2**(3), 169–172 (2006)

Chapter 6
Summary and Main Conclusions

The deposition of nanostructures by focused beams, of either electrons or ions, is a novel method of nanofabrication with potential. These deposition techniques, known as focused ion beam induced deposition (FIBID) and focused electron beam induced deposition (FEBID), have been developed in the last 30 years and are now being allocated a large number of scientific and technical resources in order to achieve many goals in the field of nanotechnology. Some of these goals, which we consider essential, are (1) understanding the physical phenomena associated with the deposition process by both experiments and complex simulations of the process. (2) Obtaining the best resolution in thickness and width. (3) Fabrication of nanostructures that present interesting physical properties of interest in micro- and nanoelectronics.

The work for this thesis has been directed toward the fabrication of nanostructures of Co, Fe, and W by FEBID/FIBID and subsequent characterizations. The Co- and Fe-based structures have potential applications in spintronics as a result of their magnetoresistance and Hall effect at moderate fields and at room temperature. W-based materials exhibit superconducting properties at moderate temperatures, exhibiting excellent qualities for basic research studies of the phenomena associated with superconductivity and applied research studies within nanotechnology.

The cooperation with the Advanced Microscopy Laboratory (LMA), Institute of Nanoscience of Aragon (INA), University of Zaragoza—a reference center in the investigation of nanosystems physics and equipped with the most advanced infrastructure and human capital—has been fundamental in the development of this thesis.

The main results presented in this thesis can be summarized as follows:

(1) Fabrication of ferromagnetic Co nanostructures by FEBID:

- Co nanostructures (>90 %) fabricated by FEBID using $Co_2(CO)_8$ as the precursor material varying several deposition parameters [1, 2]. The Co content of the deposits greatly influences their magnetic properties and thus their future applications in logic, sensors, and magnetic storage.

R. Córdoba Castillo, *Functional Nanostructures Fabricated by Focused Electron/Ion Beam Induced Deposition*, Springer Theses, DOI: 10.1007/978-3-319-02081-5_6,

- The microstructure and composition variation of nanodeposits on the nanoscale as a function of their Co concentration. Deposits of ~65 % Co composed of nanocrystals of hcp Co and oxidized Co species embedded in a carbon matrix and deposits of ~90 % Co composed of polycrystalline hcp and fcc cobalt films have been obtained [3].
- Fabrication of Co ferromagnetic nanostructures of 30 nm width, breakthrough to the functionality of the FEBID technique [4], in which the fabrication of deposits narrower than 10 nm is possible [5] as well as the fabrication of three-dimensional ferromagnetic structures [6].

(2) Fabrication of ferromagnetic Fe deposits by FEBID:

- Fe microwires grown by FEBID, using $Fe_2(CO)_9$ as a precursor material, exhibit ferromagnetic character when the Fe/O ratio is high and the Fe content is higher than 65 at %. They have a microstructure totally different from that obtained by other growth techniques, which has a major impact on the magnetotransport properties [7].
- The resistivity of the deposits scales with Fe content, and its temperature dependence ranges from a semiconductor to a dirty metal as a function of the Fe concentration. The higher Fe content samples show a resistivity nineteen times that of pure Fe and also exhibit AMR. A transition from AMR to ITMR was observed when the Fe content was decreased (by increasing the oxygen content) [7].
- Determination of a giant AHE, higher than in pure Fe microwires, was observed, which has been explained in terms of a high skew scattering caused by the complex and heterogeneous microstructure of this system [8], and future applications in new magnetic devices, such as magnetic sensors, are expected.

(3) Fabrication of superconducting W deposits by FIBID:

- Superconducting W deposits grown by FIBID, using $W(CO)_6$ as a precursor material and varying several deposition parameters.
- Correlation between the superconducting properties of W deposits to modify the lateral size of the structures, from microwires to ultranarrow NWs of 50 nm width. The critical current density increases by one order of magnitude in the ultranarrow NWs with respect to microwires of ~200 nm width owing to the energy barrier at the edges. The dissipation-free state in ultranarrow NWs was observed in a determined range of applied magnetic field and temperatures [9].
- Observation of a nonlocal voltage associated with the transverse vortex motion resulting from the applied current due to the Lorentz force, through a channel in which no current flows.

The deposition control and design of superconducting materials at the nanoscale is opening interesting fields of application in condensed matter physics, both basic studies of superconductivity and applied studies in nanotechnology having been addressed in this work.

References

1. Fernández-Pacheco, A., De Teresa, J.M., Córdoba, R., Ibarra, M.R.: Magnetotransport properties of high-quality cobalt nanowires grown by focused-electron-beam-induced deposition. J. Phys. D Appl. Phys. **42**(5), 055005 (2009)
2. Córdoba, R., Sesé, J., De Teresa, J.M., Ibarra, M.R.: High-purity cobalt nanostructures grown by focused-electron-beam-induced deposition at low current. Microelectron. Eng. **87**(5–8), 1550–1553 (2010)
3. Córdoba, R., Fernández-Pacheco, R., Fernández-Pacheco, A., Gloter, A., Magén, C., Stephan, O., Ibarra, M.R., De Teresa, J.M.: Nanoscale chemical and structural study of Co-based FEBID structures by STEM-EELS and HRTEM. Nanoscale Res. Lett. **6**, 1–6 (2011)
4. Serrano-Ramón, L., Córdoba, R., Rodriguez, L.A., Magén, C., Snoeck, E., Gatel, C., Serrano, I., Ibarra, M.R., De Teresa, J.M.: Ultrasmall functional ferromagnetic nanostructures grown by focused electron-beam-induced deposition. ACS Nano **5**(10), 7781–7787 (2011)
5. van Kouwen, L., Botman, A., Hagen, C.W.: Focused electron-beam-induced deposition of 3 nm dots in a scanning electron microscope. Nano Lett. **9**(5), 2149–2152 (2009)
6. Serrano-Ramón, L.E.: Thesis, Universidad de Zaragoza (in progress)
7. Lavrijsen, R., Córdoba, R., Schoenaker, F.J., Ellis, T.H., Barcones, B., Kohlhepp, J.T., Swagten, H.J.M., Koopmans, B., De Teresa, J.M., Magén, C., Ibarra, M.R., Trompenaars, P., Mulders, J.J.L.: Fe:O:C grown by focused-electron-beam-induced deposition: magnetic and electric properties. Nanotechnology **22**(2), 025302 (2011)
8. Córdoba, R., Lavrijsen, R., Fernández-Pacheco, A., Ibarra, M.R., Schoenaker, F., Ellis, T., Barcones-Campo, B., Kohlhepp, J.T., Swagten, H.J. M., Koopmans, B., Mulders, J.J.L., De Teresa, J.M.: Giant anomalous Hall effect in Fe-based microwires grown by focused-electron-beam-induced deposition. J. Phys. D: Appl. Phys. **45**(3), 035001 (2012)
9. Córdoba, R., Baturina, T.I., Sesé, J., Mironov, A.Y., De Teresa, J.M., Ibarra, M.R., Nasimov, D.A., Gutakovskii, A.K., Latyshev, A.V., Guillamón, I., Suderow, H., Vieira, S., Baklanov, M.R., Palacios, J.J., Vinokur, V.M.: Magnetic field-induced dissipation-free state in superconducting nanostructures. Nat. Commun. **4**, 1437 (2013)

Appendix A
Platinum Nanostructures Grown by Focused Electron/Ion Beam Induced Deposition

This appendix presents studies of volume per dose, composition, and high resolution transmission electron microscopy performed on Pt deposits grown by focused electron beam and ion beam induced deposition. The results allow us to correlate the properties studied and different conduction mechanisms proposed in the literature.

Without any doubt, the type of deposit most studied in the literature based on FEBID and FIBID techniques is platinum. The secret of its success lies predominantly in the nature of the precursor materials used for this purpose, in most cases organometallics, which are well known in chemical vapor deposition (CVD). Therefore they are easy to adapt to the technology available in a dual beam system.

We consider it very important to know the deposition rate, composition, and microstructure of the deposits in order to understand and justify the different conduction mechanisms proposed in various publications. In previous studies, the resistivity of Pt nanowires grown by FIBID and FEBID differed considerably as a function of the deposition parameters, although, in general, the deposits grown by FIBID showed lower resistivity values [1–5]. Different conduction mechanisms for Pt nanodeposits have been proposed, but a clear overall idea has not been established, mainly due to the important role played by the carbonaceous matrix. The low Pt percentage in the deposits compared to the carbon from the precursor material is the main handicap to metallic conductivity.

A.1 Platinum Nanostructures Grown by FEBID and FIBID

The precursor material used in our study was cyclopentadienyl trimethyl platinum, $(CH_3)_3Pt(C_pCH_3)$, which is solid at room temperature. This precursor was heated to 45°C. Platinum deposits with dimensions 5 μm × 5 μm were grown on a Si substrate doped with boron. The deposition parameters that were kept constant were overlap 0 %, dwell time 1 μs, refresh time 0 s, $P_{base} = 1 \times 10^{-6}$ mbar, $P_{process} = 2 \times 10^{-5}$ mbar for FEBID and for FIBID overlap 0 %, dwell time 0.2 μs, refresh

R. Córdoba Castillo, *Functional Nanostructures Fabricated by Focused Electron/Ion Beam Induced Deposition*, Springer Theses, DOI: 10.1007/978-3-319-02081-5, time 0 s, $P_{base} = 1 \times 10^{-6}$ mbar, $P_{process} = 2 \times 10^{-5}$ mbar. Other key parameters, such as accelerating voltage and electron/ion beam current, were varied in order to study their effect on the Pt content, volume per dose, and microstructure.

To calculate the volume per dose VD (μm^3/nC) of the square-shaped deposits (5 × 5 μm, we must know the thickness, obtained from cross sections, and the electron or ion dose. This was calculated for a large number of nanodeposits with different deposition parameters, which are specified below (see Fig. A.1).

The protocol was as follows: Pt deposits were grown by FEBID at several beam acceleration voltages V_e, from 1 to 30 kV, and at each fixed voltage different deposits were grown at currents I_e from 0.4 to 9.5 nA. It was observed that the electron beam current did not have a significant effect on the VD obtained, suggesting an electron-limited regime.

Figure A.2a shows VD as a function of V_e. We can see VD decreases as a function of V_e, as described for Pt deposits grown by FEBID [7, 8]. As already noted, the decline is quite pronounced above 5 kV and flattens out at around 15 kV. VD at 30 kV is one quarter that at 1 kV. This dependence is explained by the decrease in the number of secondary electrons generated in the substrate and reaching the surface of the sample [7–11].

The probability of dissociation of the precursor molecule is a maximum at energies in the range of a few tens of eV. Moreover, the secondary electron distribution generated by the incident electron beam is a maximum in the same energy range, which favors secondary electrons that contribute greatly to the dissociation of the precursor molecule. As the secondary electron yield decreases with increasing incident acceleration voltage, it is obvious that the deposition yield will decrease with increasing V_e.

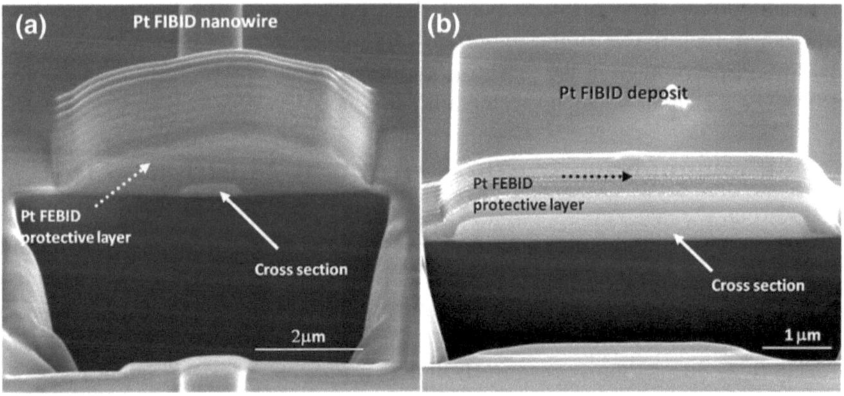

Fig. A.1 SEM images of the cross section of **a** a Pt nanowire grown by FIBID and **b** a square-shaped deposit grown by FIBID. In both cases, in order to protect the area of interest and to obtain good contrast in the image and so to be able to measure the deposit thickness of interest, before the ion milling a Pt layer grown by FEBID was deposited on top of it. Adapted from [6].

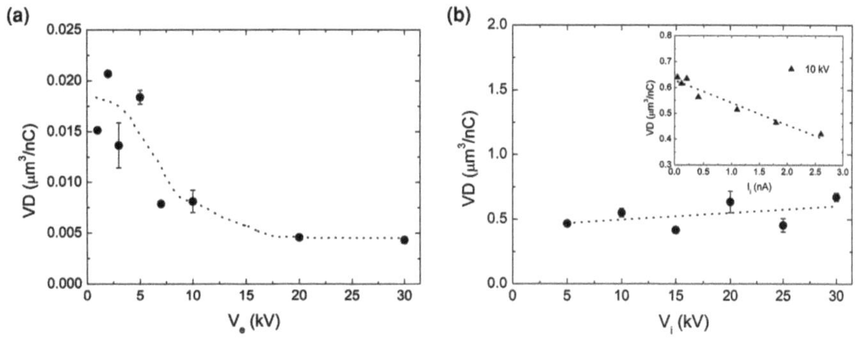

Fig. A.2 **a** Mean and standard deviation (bars) of the VD in Pt deposits grown by FEBID at all beam currents investigated as a function of V_e. **b** Mean and standard deviation of VD in Pt deposits grown by FIBID at all beam currents investigated as a function of V_i. Inset: The deposit VD grown at 10 kV by FIBID as a function of I_i. Adapted from [6].

Pt deposits were grown by FIBID at several V_i ranging from 5 to 30 kV (see Fig. A.2b). At fixed voltage, Pt deposits were grown at different I_i, ranging from 44 pA to 2.6 nA. VD increases slightly as a function of V_i. The Ga^+ FIB interaction depth, even at 30 kV, on silicon is about ~50 nm (SRIM simulations) [12], which allows the generation of a large number of secondary electrons which reach the substrate surface more easily and dissociate the precursor molecules, increasing the VD of FIBID by two orders of magnitude compared to the VD of FEBID. However, we must select the appropriate I_i in order to balance deposition and milling, otherwise we would see effects like that shown in the inset of Fig. A.2b, where VD decreases as a function of I_i.

A.2 Composition and Microstructure of Platinum Nanodeposits

Compositional analysis of the deposits was performed in situ after the deposition process, so deposits were not exposed to environmental conditions, avoiding the possibility that their composition had changed. The thickness of the nanodeposits was around ~160 nm. The V_e used to take the spectra was 10 kV. Figure A.3a presents the Pt content in the deposits grown by FEBID as a function of V_e. It is observed that Pt content decreases from 17 to 11 % as a function of increasing V_e. The C content in the deposit is dominant, at more than 80 %. This decrease in the Pt content is because as V_e increases, the number of secondary electrons (SEs) generated in the substrate and reaching the sample surface that areable to dissociate the precursor gas molecules decreases. However, the interaction of primary beam, the SEs and second-order SEs generated from BSEs electrons make up a complex scenario of the whole process. No significant changes in the deposit content as a function of electron current have been observed. It seems evident that

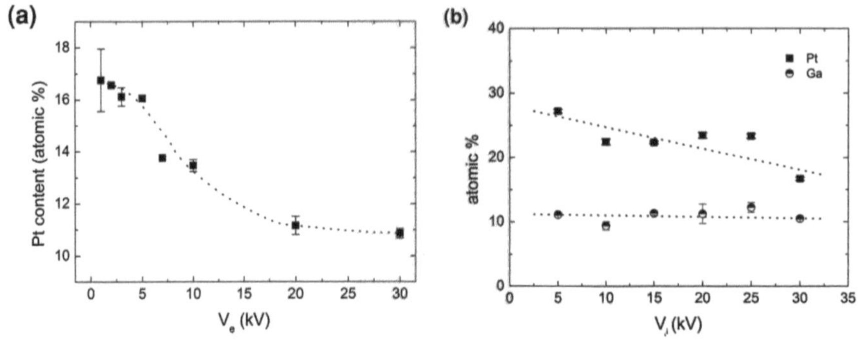

Fig. A.3 **a** Pt content and standard deviation of Pt deposits grown by FEBID at all I_e investigated as a function of V_e. **b** Composition of Pt and Ga (at %) and its standard deviation (bar) in Pt deposits grown by FIBID at all I_i investigated as a function of V_i. Adapted from [6]

the carbonaceous matrix (>80at.% in some cases) will play a crucial role in the conduction mechanism of such deposits.

Figure A.3b presents the Pt and Ga content in the deposits grown by FIBID as a function of V_i. While the Ga content is approximately constant ~10 %, it is observed that the Pt content decreases from 27 to 17 %. While the C content is still significant, more than 60 %, the metal content (Pt + Ga) is relatively high, which may be responsible for the transport properties manifested by these deposits.

The study of the deposits' microstructure was carried out using HRTEM measurements. The study was performed on Pt deposits grown directly on a commercial Cu grid with a carbon membrane transparent to secondary electrons. Other Pt deposits were studied by inspecting a lamella with thickness below ~50 nm, after the deposit had been grown on a silicon substrate. In the first approximation, we obtain clear images even though the deposits were grown on

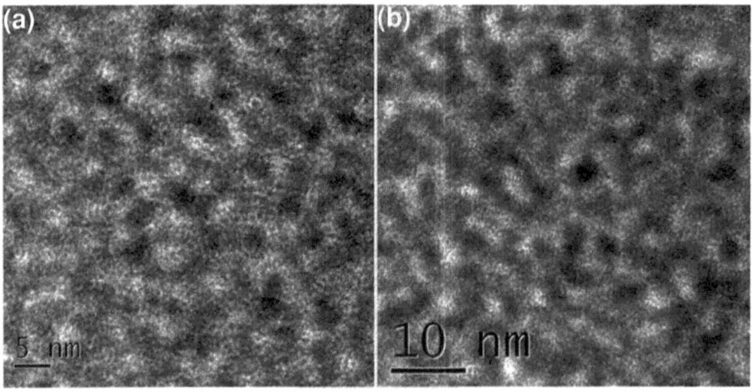

Fig. A.4 HRTEM images of Pt deposits grown at 5 kV by **a** FIBID and **b** FEBID using the dual-beam preparation method. Reprinted from [6]

different substrates. The two studies demonstrate that the two sets of results are compatible.

From EDS results, we know that the main elements of the deposits are C and Pt. However, the interpretation of HRTEM images can be tricky because we obtain two-dimensional images of three-dimensional objects, containing information from the whole thickness of the sample (50 nm). The grains are small compared to the thickness of the sample, and the transmitted electrons can traverse several grains along its path through the sample. The images obtained from the deposits

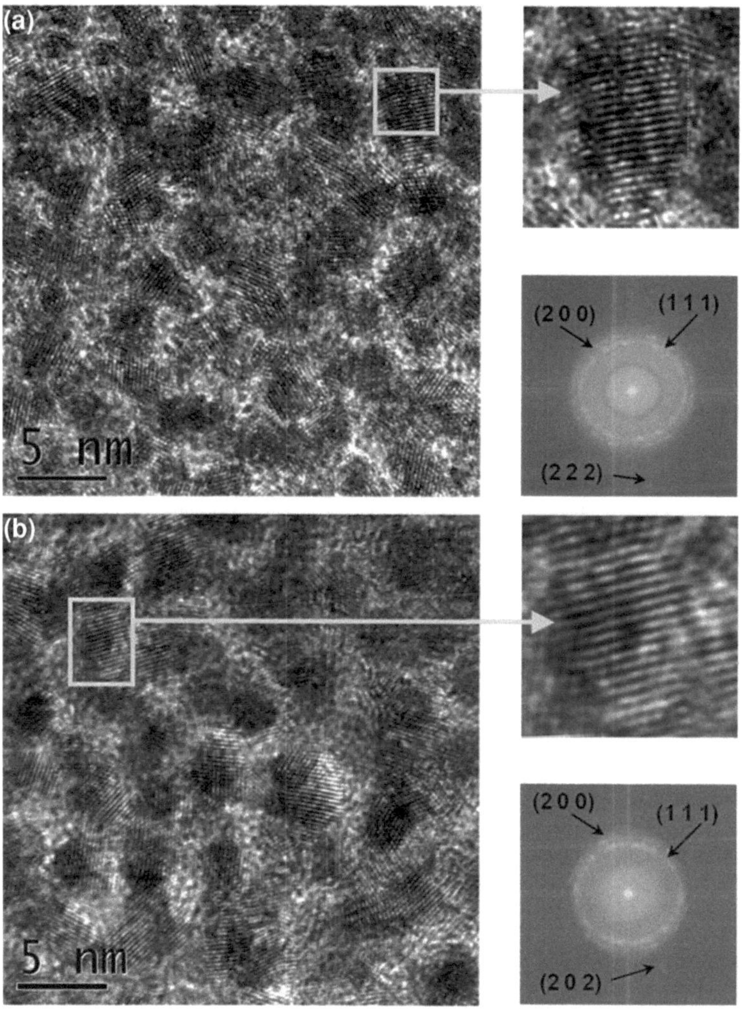

Fig. A.5 HRTEM images of Pt deposits directly grown on a commercial Cu grid at 30 kV by **a** FIBID and **b** FEBID. In each case, one of the grains was selected for clear observation of the corresponding atomic planes. The FFT of the whole image provides the diffraction spots corresponding to the atomic planes of fcc Pt. Adapted from [6]

grown by FEBID and FIBID from the lamellae are shown in Fig. A.4.

The heterogeneous distribution of Pt and C in the microstructure of the deposits has been previously published [1–5] and is explained as Pt nanoparticles with < 5nm size embedded in the amorphous carbon matrix. However, the absence of direct observation of atomic planes is inconvenient for the correct interpretation of the images. Furthermore, the observation of atomic planes in deposits grown directly on Cu grids allows more accurate interpretation of images. As seen in Fig. A.5, Pt deposits grown by FEBID and FIBID consist in ellipsoidal crystalline Pt grains embedded in an amorphous carbon matrix. From FFTs of images, Pt interplanar distances are obtained; the Pt grains have a crystalline face-centered cubic (fcc) structure with lattice parameter 0.3924 nm.

Deposits at different voltages and currents, for both FIBID and FEBID, were studied. After analyzing at least 50 individual grains in each image, we found that the average crystal size of Pt in deposits is 3.2 ± 0.8 nm.

From HRTEM images of Pt deposits grown by FEBID and FIBID, obtained from the lamellae or the deposits grown directly on Cu grids, we see that the microstructure is qualitatively similar. However, for a quantitative analysis only the deposits grown on Cu grids allow us to draw conclusions.

References

1. Langford, R.M., Wang, T.X., Ozkaya, D.: Reducing the resistivity of electron and ion beam assisted deposited Pt. Microelectron. Eng. **84**(5–8), 784–788 (2007)
2. Lin, J.F., Bird, J.P., Rotkina, L., Sergeev, A., Mitin, V.: Large effects due to electron-phonon-impurity interference in the resistivity of Pt/C-Ga composite nanowires. Appl. Phys. Lett. **84**(19), 3828–3830 (2004)
3. Rotkina, L., Lin, J.F., Bird, J.P.: Nonlinear current-voltage characteristics of Pt nanowires and nanowire transistors fabricated by electron-beam deposition. Appl. Phys. Lett. **83**(21), 4426–4428 (2003)
4. Lin, J.F., Bird, J.P., Rotkina, L., Bennett, P.A.: Classical and quantum transport in focused-ion-beam-deposited Pt nanointerconnects. Appl. Phys. Lett. **82**(5), 802–804 (2003)
5. Peñate-Quesada, L., Mitra, J., Dawson, P.: Non-linear electronic transport in Pt nanowires deposited by focused ion beam. Nanotechnology **18**(21), 215203 (2007)
6. De Teresa, J.M., Córdoba, R., Fernández-Pacheco, A., Montero, O., Strichovanec, P., Ibarra, M.R.: Origin of the difference in the resistivity of as-grown focused-ion- and focused-electron-beam-induced Pt nanodeposits. J. Nanomater. **2009**, 936863 (2009). doi:10.1155/2009/936863
7. Lipp, S., Frey, L., Lehrer, C., Demm, E., Pauthner, S., Ryssel, H.: A comparison of focused ion beam and electron beam induced deposition processes. Microelectron. Reliab. **36**(11–12), 1779–1782 (1996)
8. Takai, M., Jarupoonphol, W., Ochiai, C., Yavas, O., Park, Y.K.: Processing of vacuum microelectronic devices by focused ion and electron beams. Appl. Phys. A Mater. Sci. Process. **76**(7), 1007–1012 (2003)
9. Silvis-Cividjian, N., Hagen, C.W., Kruit, P.: Spatial resolution limits in electron-beam-induced deposition. J. Appl. Phys. **98**(8), 084905 (2005)
10. Utke, I., Hoffmann, P., Melngailis, J.: Gas-assisted focused electron beam and ion beam processing and fabrication. J. Vac. Sci. Technol. B **26**(4), 1197–1276 (2008)

11. van Dorp, W.F., Hagen, C.W.: A critical literature review of focused electron beam induced deposition. J. Appl. Phys. **104**(8), 081301 (2008)
12. Ziegler, J.F.: SRIM-2003. Nucl. Instrum. Methods Phys. Res., Sect. B **219–220**, 1027–1036 (2004)

MIX
Papier aus verantwortungsvollen Quellen
Paper from responsible sources
FSC® C105338

If you have any concerns about our products,
you can contact us on
ProductSafety@springernature.com

In case Publisher is established outside the EU,
the EU authorized representative is:
Springer Nature Customer Service Center GmbH
Europaplatz 3, 69115 Heidelberg, Germany

Printed by Libri Plureos GmbH
in Hamburg, Germany